MIMESIS
INTERNATIONAL

ATMOSPHERIC SPACES
n. 18

ATMOSPHERIC SPACES

What is an "Atmosphere"?

According to an aesthetic, phenomenological and ontological view, such a notion can be understood as a sensorial and affective quality widespread in space. It is the particular tone that determines the way one experiences her surroundings.

Air, ambiance, aura, climate, environment, genius loci, milieu, mood, numinous, lived space, Stimmung, but also Umwelt, ki, aida, Zwischen, in-between – all these words are names hiding, in fact, the founding idea of atmospheres: a vague ens or power, without visible and discrete boundaries, which we find around us and, resonating in our lived body, even involves us.

Studying atmospheres means, thus, a parte subjecti, to analyse (above all) the range of unintentional or involuntary experiences and, in particular, those experiences which emotionally "tonalise" our everyday life. A parte objecti, it means however to learn how atmospheres are intentionally (e.g. artistically, politically, socially, etc.) produced and how we can critically evaluate them, thus avoiding being easily manipulated by such feelings.

Atmospheric Spaces is a new book series whose aim is to become a point of reference for a community that works together on this philosophical and transdisciplinary subject and for all those whose research, more broadly, is involved in the so-called "affective turn" of the Social Sciences and Humanities.

Gábor Csepregi

THE PRESENT EYE PRAISES THE PRESENT OBJECT

Encountering Presences

MIMESIS
INTERNATIONAL

www.mimesisinternational.com
e-mail: info@mimesisinternational.com

Book series: *Atmospheric Spaces*, n. 18

Series Issn: 3103-4322

Isbn: 9788869775543

P.I. C.F. 0241937030

CONTENTS

To the memory of Gerd Haeffner

Introduction

Most of us would flee an unhappy, unbalanced, and unhealthy life. Intermittently, we may be inclined to produce chaos instead of harmony, misery instead of happiness, ailment instead of health. We may occasionally display a remarkable ingenuity in ignoring or casting aside what we need the most. But I doubt that men and women set out willingly and thoughtfully, day after day, in the pursuit of unhappiness and of illness. They might fail to know what exactly brings them contentment, but they still strive to attain a satisfactory state of existence.

One of the common causes of failing to achieve this pivotal objective is the dull and trifling course of everyday life, a quotidian routine of taking care of practical necessities. It can frequently cause stress, sadness and despondency. Discontent and gloom arise from the repetition of routine tasks both outside and inside home. The days are filled with the same activities, the same encounters, the same experiences; the weeks follow each other without novelty, variation, or flavour. People go through their mundane chores without any challenging surprise or annoying deviation. Monotony, boredom, and a sense of futility become the chief characteristics of their everyday life. Gradually they lose the motivation to break out from such an insipid way of living time. They also lose the ability to appreciate and produce uninhibited and spontaneous actions. Habits dull their sensibilities. Many attempt to flee from acute feeling of dullness and unhappiness by seeking excitements in alcohol, drugs, loveless sexuality, rock concert, extreme sport, gambling, or organized tourism. They thus create the impression that they have been pulled out of despair and are able to lead a more colourful existence.

The consequences of this form of existence are felt in the manner of experiencing time. When novelty and change are absent, the progress of time seems to slow down or come to a standstill: the future no longer introduces new possibilities and the past no longer brings forth accomplishments significant enough to prompt a stimulating future. Devoid of inspiring objects and actions, the present lacks density and weight and, although appears to be long and tedious, fades quickly away from memory. For this reason, time – with its barren and closed future, stale and inconsequential past, and dull present – becomes the object of attention. But the remarked time so constituted is not experienced as a genuine becoming, filled with stimulating objectives attainable in a particular world and cherished past happenings; it appears rather as a tedious and pointless duration.

What can bring a salutary change to this form of existence? How can people make their everyday life richer, more fulfilling? What can bring to their work an startling variety, astonishing flavour, and sudden surprise?

The chief thesis of this short book is that the experience of presence brings a fulfilling and enriching change to everyday life, a sort of remedy for overcoming the debilitating feeling of discontent. But let me hasten to add that encountering presences is not the only way of redeeming a sense of loss and futility. The experiences I describe in this book lend glow to a trivial thing and prompt a sense of synchrony with a surrounding. They create an atmosphere of sympathy and of understanding or even of love between two persons. They foster the enjoyment of the discovery of the expressive qualities of ordinary things and thus help to enter the world of fantasy and play. They present lifestyles, mentalities, and the more personal and more refined ways of relating to others. They induce a delightful feeling of being free from concerns and compromises. They may even create a life-long attachment to a profession or field of study. During certain unexpected moments, they allow us to know that we are alive and that we are able to enjoy the sensual delight of being alive. They fill the heart with a sense of wonder and gratitude.

In her novel *Pilgrim at Tinker Creek,* Annie Dillard wrote beautifully about her experience of presence. She noticed how this gratifying experience modifies the manner of living the present. A "great door opens on the present, illuminates it as with a multitude of flashing torches."[1] While attentively listening to a musical performance or contemplating a blossoming tree, we may also experience the extension and density of the present. Our senses dwell and concentrate on the continuing presence of tones and forms and thus make us take a footing in the present. We might rightly consider the magnetic presence of a thing as the cause of adopting the act of "letting go" of both what has past and what is ahead of us. Quite often, however, the presence of a thing and the act of paying attention to lived present create and balance each other, even though the reality of what is present has an ontological priority over the act of focusing on the present.[2] We must sense the forceful presence of a thing so that we can free ourselves from any desire of thinking back to the past or of hastening the arrival of the future thus allowing ourselves to remain fully in the immediate now. But, similarly, we must free ourselves from the obsessive preoccupation with the future or the past so that a thing could assert itself as a presence.

To put it in another way, the presence requires from our part the readiness of being present. But this openness does not guarantee a presence. It would be pointless to hope that a landscape or a child make their presence appear just because we adopt an attitude of receptivity. Presences often emerge abruptly. Their disclosure is unpredictable. They are welcomed as presents and remembered with a sense of gratitude. They occur when we realize that a thing has its own tongue, speaks to us, and communicates its being there for us. Still, we perceive and enjoy the unforeseen appearing of presence, if we relate to our environment with sensibility and attentiveness. The person who is unable to stop and dwell attentively on things will fail to become aware of the intense calling of a presence. Presences will not touch the person who, while walking on a nature trail or

1 Dillard, *Pilgrim at Tinker Creek*, p. 80.
2 Mehl, "Structure philosophique de la Notion de Présence," p. 171.

visiting a museum, stares at his mobile device. Such a person has decided not to be present, not to "catch the grace" when something – let's say an old manuscript on display at the Library of Congress in Washington, D.C. – without warning, is ready to press on him, to grip on him, and to make itself felt as a presence.

The reader will see that I have illustrated my points with poems or excerpts taken from literary works.[3] Many poems arise from experiences of presence. Persons with poetic temper have the gift of exposing under new light the common and the familiar. They catch the radiance of what often seems to us drab and insignificant. They describe, for instance, how the beauty of a valley or of a fine gesture compels attention and elicits wonder. Trivial-seeming details may come forth as a meaningful presence and the poets are able to apprehend them and express them with an "emotion recollected in tranquillity" (Wordsworth). We turn to poets because we realize that the ordinary and immediate facts of everyday life moved them and that they were able to translate their perceptions and feelings into an appealing poetic language. Although they obey the dictates of the language they use, the poets appearing in this book display an unwavering fidelity to the real. Their attachment to their lived experience driven not merely by their intellect and by their senses, but chiefly by their most personal emotions. To gaze and marvel at the presence of a thing, of an action, or of the beauty of a morning and make this vision into poetry, the experience must somehow touch the poet's most intimate being.

Ordinary persons who experience a presence often declare that it is so hard to recount verbally a feeling, a perception. It may happen that even poetic language fails to describe all aspects of a particular experience.

Yet poets teach us to see the extraordinary in the ordinary, the mystery in the familiar, the life in the unliving things. They call to our attention the forms, shapes, silhouettes, shadows, motions, and

3 I wish to express my profound gratitude to Professor Paul D. Morris for calling to my attention poems unknown to me and for making valuable recommendations for the improvement of my manuscript.

colours that we usually neglect or fail to see: the afternoon sunlight upon the living room's wall or, as in the poem of Gerard Manley Hopkins, *The Windhover*, the flight of a falcon in the morning just as the sun is rising.

> I caught this morning morning's minion, king-
> dom of daylight's dauphin, dapple-dawn-drawn Falcon, in his riding
> Of the rolling level underneath him steady air, and striding
> High there, how he rung upon the rein of a wimpling wing
> In his ecstasy! then off, off forth on swing,
> As a skate's heel sweeps smooth on a bow-bend: the hurl and gliding
> Rebuffed the big wind. My heart in hiding
> Stirred for a bird, – the achieve of, the mastery of the thing!

This book is dedicated to the memory of Gerd Haeffner as a token of my profound esteem and gratitude. I met him in 1976 in Quebec City, at Laval University. Thomas De Koninck, then dean of the Faculty of Philosophy, invited him for a semester and introduced us to each other. He was teaching philosophy as a visiting professor and I was studying theology as a graduate student. I showed Haeffner the notable landmarks of the city and invited him to my home. Until his death in 2016, we treasured and renewed our friendship through exchanging letters and meeting at the *Hochschule für Philosophie* in Munich, where he was professor of philosophical anthropology.

During his studies of philosophy, in Pullach in the Isar Valley, as a young Jesuit, Haeffner had an experience, which left a lasting effect on his thinking. In a short text, entitled *Wie ich zu meinen Themen kam. Eine philosophische Mini-Biographie*, he gave of this experience the following account:

> It had to do with one of the white-painted doors in the long corridors of our college building, through which I often passed thoughtlessly, but which one afternoon struck me as an unnatural "reality" or "presence." On this afternoon, it looked essentially no different from its usual appearance. But whereas I usually walked past it without paying it any further attention, its presence now struck me as something remarkable and fascinating. Self-critically, I asked my spiritual advisor whether this experience might have been something psychopathological. He reassured me. I then asked myself: could it be that I was experiencing the

> "being" about which Johannes B. Lotz, my teacher in the "Ontology" class spoke?[4]

Subsequent to those earliest years, Haeffner continuously reflected and wrote about the human experience of time, of living in the present, and encountering a presence. In the singular experience of presence, we come to the astonishing realization that things are and that they assert their being in an intensive and forceful manner. He pointed out that the experience of presence is a grace and, as such, is not granted on whims or fancies. Presences can emerge suddenly in the most prosaic context. We all are able to see and savour them if we adopt an attitude of attentive receptiveness. Presences open the door to the present, and our readiness to dwell in the present fosters further possibilities of welcoming presences. Furthermore, for Haeffner, the encounter with a presence serves as a ground for experiencing the divine presence.

His writings, as well as the numerous conversations I had with him over the years, influenced and inspired my own thinking. This book seeks to expand on his ideas on the objective and subjective poles of presence experiences; it highlights concrete instances of presences in our surroundings, in our contacts with things and persons and in ourselves. I hope that the examples I give make more palpable the wonder of experiencing a presence and of the astonishing awareness of the pure present. The book invites the readers to enhance their own sensitivity and attentiveness and thus to draw happiness, joy, and freedom from the concrete presences, wherever they come into their view.

4 https://hfph.de/files/2025/01/2009-philos-minibiographie.pdf

In his *A Letter to a Hostage*, Antoine de Saint-Exupéry provides the reader with a description of an experience of presence. He recounts how a person and his surroundings become a tangible presence through the exchange of smiles and the subsequent softening of faces. Seemingly unimportant and transient, the sudden brightening of faces appears nevertheless essential, arousing in both persons the awareness of trust, sympathy, and peace.

Arrested by revolutionary anarchists during the Spanish Civil War, the writer found himself in a cellar, in which reigned indifference, silence, and boredom. He expected that the guards would shoot him promptly and without any pang of conscience. Then, wanting to introduce a change into this miserable atmosphere and thus find an escape from the thought of imminent execution, he looked at one of his guards with a vague smile. He asked for a cigarette. To his great surprise, the guard raised his eyes in the direction of his face and looked at him with a faint smile.

> That smile saved me … It opened a new era. Nothing had changed, and everything had changed. The table of scattered with papers became alive. The kerosene lamp became alive. The walls were alive. The lethargy which seeped from the dead objects in this cellar was magically alive. It was an invisible bloodstream had begun to flow again through a body, renewing all the parts and restoring their meaning.
>
> The men had not moved either, but, where a second earlier they had seemed further removed from me than some antediluvian species, now they seemed reborn to contemporary life. I experienced an extraordinary sensation of presence. That is what it was: presence! And I felt kinship with them.[1]

1 Saint-Exupéry, *A Letter to a Hostage*, p. 44.

In this warmer atmosphere, established by the mutual smiles, the guard is no longer seen under the abstract category of militant anarchist, nor is the captive writer treated as if he were an unwanted and suspicious loiter. Both now appear with their distinct personhood, with their own vulnerability and uncertainty. The smiles create a human radiance, an atmosphere of understanding, a tacit willingness to ease discomfort and hostility, a genuine presence.

The exchange of smiles transforms the appearance of the surrounding objects by endowing them with life and meaning. They become more vivid, full of life and are even imbued with glow and beauty.

A presence is experienced in an encounter which stands out from the multiplicity of contacts with things and persons. Whatever has been accomplished or perceived before is left behind; it loses, for the time being, its relevance. Even if the object was previously familiar, it appears now as if it were being seen for the first time and as if it had never been part of one's surroundings.

Presence experiences often occur surprisingly even though we might open ourselves to a presence by adopting a particular attitude of openness and of readiness to respond to the call of a reality. The various accounts of presence experiences presented by Timothy Beardsworth in his book make use of the words "suddenly" and "sudden" quite frequently. A presence may be conveyed "out of the blue."[2] Although the experiences are usually short, their memory may last much longer. We may return to the immediate past with an acute feeling of regret. Or we may prefer recalling the experience with a profound sense of gratitude. We then focus not so much on the what happened in the past but on the enrichment the experience has brought to the present: we acquire a new way of seeing and appreciating the surroundings, the things around us, the other person and ourselves.

An encounter, as opposed to a meeting, which is usually the outcome of our planning process, happens to us in the present. Many things happen as we go on with our everyday lives: an illness, an

2 Beardsworth, *A Sense of Presence*, pp. 56-7.

accident, a demand issued by an unfriendly ticket controller, as well as a smile of a fellow passenger on the subway or a bright sunny morning. These happenings are either pleasant or unpleasant. They concern us, affect how we act, feel, and make our choices. Either happenings occur as we act (a violin string breaks while playing a piece) or emerge independently from our action (rain starts to fall as we stand at the bus station). Presences are such happenings: they result from our actions or merely come about while we are resting quietly. While an encounter happens, the modes of contact with something change from distance to intimacy or from being intensely close to being far apart.

A presence introduces a break into the temporal progress of the person; an activity either comes to a complete stop or receives a new meaning and a new orientation. The habits and conventions that usually guide a behaviour lose their guiding role. As the sensory qualities became captivating and immobilizing, the tasks to be achieved lose their urgency. Feelings of tension, gloom, and nervousness are relieved and give way to a sense of peace, well-being, and cheerfulness. The break can be so liberating that the experience may compel a deeper reflection on the purpose of one's existence.

Encountering presences cannot be prepared in advance. We may anticipate encountering a role model whom we hold in high esteem. Perhaps we foresee being captivated by his words. However, our hope will be deceived if the person reveals nothing about himself and, in fact, remains throughout the conversation absent. Unfortunate circumstances – noisy tourists, inclement weather, worries about finding the return path – also thwart the desire to savour the enchanting presence of a landscape. The more carefully we strive to create the encounter with a presence the more our effort fails to meet the desired result. Presences are contingent and surprising experiences; they often occur when we expect them the least. The careful staging of an artistic performance or of a religious celebration will not guarantee an experience of presence.

Yet we may create a favourable condition for such an experience. The presence may appear when we introduce a pause into our daily occupation and become disposed to catching sight of it, notwithstanding its closeness and familiarity. The object is no longer taken for granted, neglected, or forgotten. Suddenly the reality becomes the source of wonder: it is there, standing out on its own, distinct from everything else. In the words of Gerd Haeffner, "they are *themselves* something" (*Sie sind* selbst *etwas*).[3] It is there in front of us not because of our curiosity or our desire to make a use of it, but in its own right, independently of any kind of attitude we may adopt towards it, any sort of objective we would like to reach. As detached observers, we may feel privileged and delighted to share space with it, to dwell on it and to perceive its being in itself.

Nevertheless, we are also impelled to turn to the object and request from it some revelation. What is this object, asserting its presence, able to tell us?

The object may convey a value, a virtue, an attitude, or a cherished memory. Wisdom, alacrity, or inspiration may come forth in the movements or words of a revered actor. In the teacher's attitude, students are able apprehend the passion for scientific investigation and the beauty of the subject presented in class. Similarly, tenderness, recognition, or love are detected in a gift. These are felt but, obviously, remain unseen. Exercise books used during former school years reveal sweet memories or sour disappointments. As we take them into our hands and flip through the pages, we see all the works and disciplined hours we have put into our notes or essays. It is endowed with the learning experiences of our youth, which we cherish to preserve and keep somehow alive. In all these examples, as well as in the following one, mentioned by Georges Haldas, a presence brings forth the unity of visible and invisible:

> You have before you a being to whom you speak and who speaks to you. (…) you have before you his face, with its full range of expressions, his body, his gestures, those of his hands in particular—in short, the visible part of them. But, at the same time, hidden from your eyes, lies his entire history, everything he has experienced over the years, everything that has nourished, so to speak, his inner life and

3 Haeffner, "Präsenz,"p. 132.

> his maturation: the invisible part of his being. Now, what exactly is happening in this instance? Whether he speaks or remains silent, a subtle flow of current and wave emanates from this being, and reaches your solar plexus. This constitutes his presence.[4]

An enchanting moment may be recalled with the painful regret of not being able to relive it again. But our memory of a deeply moving experience may induce us to see things with different eyes and to become receptive, in an everyday context, to a sudden appearance of a presence. In this case, remembering presences of the past helps us to remain open to the eventual presences in the present.

The presence often brings forth more or less important memories. Happy or painful events or encounters are relieved, once again, while contemplating a scene or a thing. Unrealized plans and desires come forth while meeting an old acquaintance or while reading a letter from distant times. The conversation or message may be banal, but the spoken or written words assert their sharper tonality because they point to a segment of the past.

Let us return to the experiences occurring in the present. In a smile, as we have already noted, we detect a certain animation, an expressive quality. A smile may express the embarrassment of someone making a blunder. It may divulge the disapproval or contempt for someone's behaviour or choice. But a smile may radiate, illuminate the whole face, and reveal the character of the person, an inner desire, a wish to say something and to request a reply.

Similarly, we notice, while contemplating a landscape, a vibrant presence. We sense a power or peace in a hill or a countryside and this quality defines the whole atmosphere arising from a natural scene. This atmosphere puts us in the presence of serenity and amenability. However, during a somber day, we come across the strangeness and defiance of the same landscape.

The apparent will or caprice of the object to tell us something makes us stop, admire, contemplate all its aspects with awe. The presence may have little connection with a lived situation. While

---

4 Haldas, "Les sept piliers de l'état de poésie," p. 121.

looking at the bright colour of apples or cherries, we are stirred by a noisy market place. But we often need solitude, recollection, and above all stillness to perceive hills and meadows with a sense of exhilaration.

A building may elicit the same experience. One day, I walked alone into a small stone church that has been, for years, out of use. From the outside, the church did not evoke any distinct impression. But, once finding myself inside, the altar, the walls, some discoloured paintings, the old benches, the silvery organ pipes above the narrow choir loft called me to stay. They irradiated life of their own. I became open to their suggestion, to their vitality, to their presence. What is in the church or in another reality that made me stay for a while, to look around with an attentive eye, and to live fully in the present?

According to Gerd Haeffner, to come to a singular experience of presence, we must go through a number of steps, as it were through a personal "road of discovery."[5]

First: we perceive a reality – a tree in the park or a river running through a town –in the present, in such a manner that we are no longer constrained by the future and we do not feel the distracting weight of the past. Yet the future and the past do not vanish completely. We are able to pay attention fully and intensely to the tree or the river as they appear in the present.

Second: our attention becomes centred because we are drawn to an object. In fact, it is the object that seizes us. We have the feeling that the object is opening up to us, even looking at us and telling us something. We find it communicative and meaningful. This feeling of being touched by such an openness holds us to the present and makes us dwell spellbound in front of the object. I, as a subject, am involved in the perception and thus contribute to a certain extent to the quality of the impact of what is perceived.

Third: we become conscious of the fact that we are in relation to an object and to its presence with a sense of unity. Its presence appears in a certain manner and with a certain degree of intensity. It cannot be reduced to either material or psychological facts, nor

5 Haeffner, "Präsenz," pp. 133-4.

to both together. I would call it an intense radiation of an invisible affective force that exerts an effect on us, as we perceive the object with heightened attention. The presence is an irreducible, primary reality, emanating between the object and ourselves. We sense it with delight and amazement and dwell on it for a while.

Let us look at these stages in a more concrete way. Many of us, living in large cities, may share Max Picard's observation made in the warn-torn Milan.[6] While walking leisurely on a teeming street, we see people advancing towards us and heading to the opposite direction. We perceive their empty and unresponsive faces. They seem all the same: they are uniformly inexpressive and spiritless, perhaps formed by previous deceptions or defeats. Suddenly we notice, among all these apathetic walkers, a glowing and unscathed face. In contrast to the colourless appearance of the other people, this face appears with sharper features, radiant eyes, and calm lips. The face is calling on us to pay attention, to slow down our pace and to linger intently on this perception. We open ourselves with awe and contentment to this singular appealing face. It is there in front of us in all its concreteness and we become aware of both its being-of-its own (*Eigen-Sein*) and its being-for-us. We also realize that, to our astonishment, only this distinctive face alone is *there*; it enthrals us with its telling, vivid, and irradiating *presence*. All the other faces seem unreal, virtual, as if they were merely images projected on a screen.

This happy and fortuitous encounter with a stunning face brings into the fore the satisfaction that we feel when we find ourselves with someone in a relation of synchrony. We feel ourselves in communion with the other person; we are one with her. The sense of oneness with another person is a basic feature of the sense of presence.[7] But experiencing a relation of synchrony also makes us painfully aware of its absence: desynchrony remains a central part of human existence. Opportunities, encounters, recognitions happen when we are no longer able to enjoy them. Proficiency blooms when

6 Picard, *Zerstörte und Unzerstörbare Welt*, pp. 10-11.
7 Beardsworth, *A Sense of Presence*, pp. 59-62.

there are no longer occasions to make a good use of it. Sympathetic understanding occurs when the person we finally understand is no longer present in our lives. But, as we will see later, objects and persons are still able to assert their presence long after they have lost their relevance. We are better able to see and recognize them than while they were part of our daily lives.

Painters, sculptors, or poets are perhaps the most receptive persons for the encounter with such a presence. They are waiting to be seized by a movement, an attitude, an unusual gesture. Once they are back at their workshop or beside their desk, they are eager to create a figure or to write a poem. The poet Rainer Maria Rilke saw in the perception of exceptional presence a rare blessing for the artist:

> There are, in fact, moments when a person stands out from his splendour in clarity and silence before you. These are rare festive pleasures that you never forget. You love this person from then on. In other words, you make an effort to retrace with your own tender hands the outlines of the personality that you came to know in this hour.[8]

As I have said, a presence comes to us all of a sudden. It sends a magnetic call. If we feel an affinity with its form and the particular content this form suggests, we are eager to respond. We give an answer to the object with our whole body; our hands and legs make the appropriate movements to reach for it or to preserve a respectful distance over and against it. We are both active and passive: we make movements in order to attain a goal and we are moved by the presence of a reality. It happens that, when we expect it the least, a seemingly trivial thing triggers a motor response executed with ease and concentration. We have the impression that we are carried by a source of energy and spontaneity concealed somewhere within our body that we do not need to control and that initiates familiar actions or introduces new movements, and even responds appropriately to the demands and challenges of the situation. We may then come to a presence, which is not an external reality, but the concentration and virtuosity of our own body.

8 Rilke, *Notizen zur Melodie der Dinge*, VII.

Presences are freed from the concerns of the future and the weights of the past. The progress of the lived time seems to stop and the past or the future seem to offer their blending support for the experience of pure present. The clear distinction between the temporal segments – past, present, future – seems to fade away as the presence of an object grabs our full attention. Such a presence is, for example, that of a yellow field of ripe grain, seen under blue sky, at noontime, during a hot summer day.

In his novel *Stoner*, John Williams described how the reciting of a Shakespeare's sonnet and the subsequent questions raised by a teacher about its meaning introduced a radical change into a student's life. When the teacher recited the sonnet and spoke to the student, this encounter lifted the student out of the flow of time. The student had become "someone other than who he had been."[9] Later on, the reading of a literary work in his attic room at night provided him with an intense experience of dense presence.

> And he would feel that he was out of time, as he had felt that day in class when Archer Sloane had spoken to him. The past gathered out of the darkness where it stayed, and the dead raised themselves to live before him; and the past and the dead flowed into the present among the alive; so that he had for an intense instant a vision of denseness into which he was compacted and from which he could not escape, and had no wish to escape.[10]

Although the experience implies a delightful feeling of the loss of the sense of passage of time, the unconstrained and dense present cannot be extended indefinitely. Such experiences of stepping out of the flow of time are usually brief. We remain anchored in the present without trying to lengthen it. The feeling of intensity comes precisely from the brevity of the experience and our unwillingness to be concerned about the duration of present. Wanting to extend the time of duration with eyes fixed on a watch results in the dissipation of the experience of pure present.

---

9 Williams, *Stoner*, p. 16.
10 Ibid.

Annie Dillard has sensitively described her experience of the presence of a scenery seen around an ordinary gas station.[11] Her experience was intense and brief. Whether she wanted to extend the duration of the experience or end it prematurely, her will did not matter. "The present is an invisible electron; its lighting path traced faintly on a blackened screen is fleet, and fleeing, and gone."[12]

There are "certain inexplicable moments" when, just like Annie Dillard, we "catch grace" and become aware of a presence intimating peace and harmony.[13] We have then the impression that all is well and our existence is full of meaning and promise of fulfilment.

We experience such a moment, in particular, when we sit quietly in our room, at the twilight of the day, and remark the rays of sunshine illuminating the wall and the objects around us. We take a pleasure in their splendid luminosity and sense in them order and stability. We feel the gentleness of the sun on our body, as we begin to forget our troubles as well as the urgency of our daily tasks. In these moments, Romano Guardini tells us, the object of a presence allows us to catch a glimpse of "some ultimate and unique reality. A mysterious presence, to which we can give no name, radiates contentment."[14] Once again, this experience of presence cannot be brought on command or lengthened at will. The presence of mystery often comes as a favour, as a grace while we find ourselves in a state of solitude and stillness.

In his novel *The Moviegoer*, Walker Percy relates the memories of Binx Bollings who, unlike his friend, was all of the sudden affected by the mystery and wonder of a summer afternoon:

> The August sunlight came streaming in the great dusty fanlights and lay in yellow bars across the room… In the course of an afternoon the yellow sunlight moved across old group pictures of the biology faculty. I became bewitched by the presence of the building; for minutes at a stretch I sat on the floor and watched the motes rise and fall in the

---

11 Dillard, *Pilgrim at Tinker Creek*, pp. 77-81.
12 Ibid., p. 80.
13 Ibid., pp. 81-2.
14 Guardini, *Freedom, Grace and Destiny*, pp. 92-3.

> sunlight. I called Harry's attention to the presence but he shrugged and went on with his work. He was absolutely unaffected by the singularities of time and space.[15]

Natural scenes may also become illuminated all around. Bushes glow a radiance and bright light is streaming in through the trees. We perceive them as singular sources of presence experiences. Seen and felt everywhere around, the light seems to have no source of emanation. Reality is lit up from inside. Our gaze meets "the invigorating rays of reality."[16]As the vague and blurry representations disappear, everything becomes clear and definite. Colours are more brilliant, lines more pronounced, sounds sharper and clearer. The experience conveys a sense of warmth and contentment, sometimes an exquisite feeling of bliss and serenity.

The French classical scholar Jacqueline de Romilly gave this enthusiastic description of the contemplation of light and shades:

> Overwhelmed with worries, squeezed into the subway of our overcrowded cities, our eyes on the newspaper, haste in our hearts, we forget the light. But each time I come to my house in the Province, I feel the same amazement again before the play of light and shadow. When I see, in the morning, a section of wall suddenly light up in a golden blond, and the shadow of the plane trees moves gently, in subtle patches, against this luminous background, or when I see the green of the pines all caressed by the sun, and the jagged slope of the Sainte-Victoire mountain, with its whiteness changing reflections at every hour, I have the feeling that nothing, ever, can surpass the joy of this contemplation.[17]

---

15 Percy, *The Moviegoer*, pp. 51-2.

16 Guardini, *Les signes sacrés*, p. 22. The expression is from the preface of the first edition of *Von heiligen Zeichen*, translated to French. This preface is unfortunately absent from the more recent German edition, as well as from the English translation of this work. Guardini tells us in this preface that only the concrete contact with reality helps us to renew our lives. We need to feel "the shock of things."

17 De Romilly, *Ce que je crois*, p. 17.

She adds that one does not need to travel to Provence to witness such a presence. Patches of sunlight on the green lawn can be seen in every public park.

Although we may sense a mysterious presence in the silence of a sunny afternoon, most our experiences of presence put us in touch with persons or tangible realities. When we meet a person and this person becomes a presence, we go beyond roles, titles, and general designations. Whether we encounter a person or a thing, their presence is always concrete and singular. An abstract idea, principle, norm, value are not given to us in the form of presence.

The posture, gestures, and movements of a person give us access to a particular emotional quality, an atmosphere. The personal atmosphere comes through a glance, a vibration of the voice, a gesture of the hand, an exquisite refinement of manners; they all disclose a marked character. Teachers or artists impress us with their strong personalities. They affect all those who find themselves in their presence. In spite of their lack of striking physical traits, they exert a distinctively vivid impression: they radiate inner depth and conviction.

In every presence, there is an enigmatic self-giving. A teacher makes herself present and remains forever as a determining influence in the memory of her students. Her presence presupposes the willingness to impart something of herself to the students and to display, in her teaching, passion, depth, and generosity. A presence as a generous giving and even self-giving can be welcomed or refused; it cannot be produced by an explicit demand. It is impossible to teach and order someone to create a presence or make oneself present or, conversely, to show receptivity to a presence.

If two persons give themselves to each other and, in exceptional moments, perceive each other as a presence, this free donation of themselves is never without some restraint; they always hold back something of themselves even if this reserve may give rise to misunderstandings. Lovers or friends manifest to each other as "pockets of *resistance*."[18] In absence of such a reserve, they risk relating to a distorted image created for a certain purpose.

---

18 Lewis, *The Four Loves*, p. 75.

The act of self-giving is equally manifest in things that we use or with which we play in our surroundings. An old table, hosting so many family gatherings, seems to come near a person and to send an invitation to touch its surface and to cherish its durability. Similarly, the water fountain pulls the child closer. The water spurting swiftly enthrals him. The child has the impression of answering back to a call coming from a living being. He joyfully accepts the invitation. The water fountain or any other inviting object tells him something. He may then even identify himself with unliving or living realities. Sometimes the nearness and pressure of a thing is strongly felt even if it is not seen. One is aware of the proximity of a friend, his "affective force" (Beardsworth), even if this person is not visible. This sense of nearness, this genuine *sense of presence* may lead to the feeling of complete closeness, oneness with the other person.

Presences emerging in a human encounter are often elusive. The presence of a fine act of curtesy is usually fleeting, transient. We are unable to notice, in the present, with precision all aspects of its form and the extent of its significance. Time and again, the appearance and meaning of a gesture of encouragement or of recognition are fully grasped in retrospect. A gentle embrace conveys warmth and affection but in the moment of trying to put into words what we experience, its presence is already lost. We merely remember it, as unforgettable gesture, with a sense of gratitude.

In his thorough study, Roger Mehl reminds us of the relation between the presence and the act of giving a present, a gift.[19] The presence of someone who gives is eminently manifest in a present offered to a fellow human being out of sincerity. The person gives himself in the present: his love, gratitude, esteem, care, respect, solicitude. A book or a box of truffles conceal all sorts of feelings or thoughts. The present is genuine if, as a presence, it is the sign of an intention which animates it and if the person receives it without any intermediary reality. Obviously, sending and receiving the photographic image of flower via mobile phone conveys no presence. In the absence of placing the present into the hands of

19 Mehl, "Structure philosophique de la Notion de Présence," pp. 174-5.

another person, the object would lack the presence of the person who gives.

The present elicits a sense of gratitude and a desire to give something back in return. In a similar manner, the presence touching one's whole existence may provoke an appropriate reaction. The beauty of a mathematical truth may create an enduring interest in a scientific domain. A well-recited poem, in which a gifted actor or a passionate teacher is compellingly present, may prompt a life-long interest in arts. The encounter with of a sick person lying on a hospital bed may change one's outlook on life. Whether experienced in a concert hall, in a school, or at home, the presence of a song or a piano sonata may produce a lasting passion for a musical profession. The Hungarian composer Zoltán Kodály recalled not only his life-defining early musical experience, but also the presence of a persistent concordance between an aural and a visual impression:

> One event remains unforgettable for me: in the afternoon, the setting sun bathed the whole house as if in flames, my parents were playing music, and that piece of music is still associated for me, to this day, with that illumination; only much later did I realize that it was a sonata by Mozart that my parents had played.[20]

As a contribution taken from Timothy Beardsworth's book emphasizes, the plain vision of a red-coloured sky during a late summer afternoon may also alter a whole outlook on life.[21] In all these experiences, the aroused emotions influences the decision about the course of life one will pursue.

As I have pointed out, presences may bring memorable experiences or even life altering changes into our existence. But they do not come to us only in uncommon settings. They are all around us. In numerous everyday and prosaic situations, as long as we keep our eyes open and adopt an open and receptive attitude, we are able to experience astonishing and persuasive presences. In the following pages, I will examine some aspects of presences that surprise us as

20 Kodály, *Mein Weg zur Musik*, p. 20.
21 Beardsworth, *A Sense of Presence*, p. 41.

we move around in our surroundings, as we pause and see things attentively, or as we encounter persons and become aware of their salutary influence on us or, "far from the madding crowd" (Thomas Hardy), as we collect ourselves and overcome distraction in the present.

For my analysis of the presence of a landscape, I derive my initial inspiration from the entrancing travel notes of Antal Szerb published first in 1936. Szerb, a brilliant Hungarian writer and scholar, experienced the vibrant and exhilarating presence of the serene Italian countryside when, during a warm and sunny summer afternoon, he reached the Montale Tower, located on the third peak of the Monte Titiano, in San Marino.[1]

> I sit at the foot of the tower and gaze out at the view. My view. So far I have been obliged to share it with the day-trippers, to join enthusiastically in their constant chattering. Now I have come into possession of my soul. The whole of this part of the country is mine: on one side, rich, twilit Romagna, with its scattering towns, sloping gently down to the distant sea; on the other, the bandit-haunted Apennines of ancient Etruria. Behind them again, I sense the presence of my Easter Kingdom: Urbino, Arezzo, Gubbio, and the whole of Umbria. These are real mountains, as vast as a man could wish. Mountain any larger than these I do not like. I dislike the Alps. They are unrestrained, titanic. In a word, inhuman. No mountain should be greater than these before me. The Apennines are human-like, just as the entire Italian landscape is human-like. And this is why it is more beautiful than any other landscape of the world.[2]
>
> I sit looking out over the Italian countryside. In the advancing twilight, the blues and the reds on the skyline will be sharper, more sonorous,

1 Szerb, *The Third Tower*, p. 68.

2 I have translated the Hungarian "*emberszabású*" with the English "human-like" and not with "human-scale," which is proposed by the translator of the book. The expression "human-scale" refers to the abstract geographical environment. It is measured by distances and positions.

> more distant; I absorb the inexpressible sweet serenity of the Italian landscape, and for the first time on my present journey, I am happy. Happy in the archaic sense of the world, according to which no child can be said to be happy: a replete happiness, containing everything.

A landscape is present through a number of material realities: mountains, hills, trees, rocks, meadows enamelled with flowers. Our senses perceive them with all their tangible characteristics. First, we perceive their outlines and then, upon careful observation, we discover the details. The material presence of a landscape is unpredictable; a tree, a stone, a creek may suddenly appear here and there following our slowly moving perusal of the overall scene. A landscape is prone to continuous metamorphosis; it gives itself in different ways; it lends itself to the variation of the light, of the season, of the whims of the weather, as well as to the changing standpoint and chosen perspective of the beholder.

Unlike a work of art, which presents a winter landscape in a certain frame, a snowy field apprehended as a lived-experience, gives itself with no well-defined edges. Its outline is not stable. It extends and shrinks as we move from one place to another and as the panorama modifies itself in accordance with the changes to our location. A landscape "keeps improvising," presenting new aspects and appearances of the very same material reality. The modification is produced by the change of our standpoints and bodily movements as well as by the inevitable metamorphosis of the external conditions and circumstances.

The landscape and our body's movements and aptitudes mutually enhance each other. We touch the landscape, taste it, and fully embrace it with our senses and, by paying close attention to it, we press the landscape to reveal its hitherto hidden aspects.

Whether we sit calmly in order to contemplate a lake, walk on a path between the trees of a forest, or climb a steep slope, we are bodily present to a landscape. The landscape always affects our body in a certain way and our bodily attitudes and movements reveal a particular aspect of the landscape. The features of the landscape may then encourage leisurely walks, challenging hikes, or invigorating swims. Indeed, the landscape solicits our body and we respond to this solicitation by adopting a particular bodily attitude (standing, sitting, or lying down), by accomplishing bodily movements (walking, running, climbing, or swimming), and making ingenious use of our bodily aptitudes (sensibility, spontaneity, memory, or imagination).

In his reflections on the immediate sensory communication with appearances and on the detached understanding of things and people, Erwin W. Straus makes a clear distinction between the landscape and geographical space.[3] In the landscape, as we move from one place to another, we respond to the sensory impressions coming to us from various locations. We are always in the middle of the landscape and, in this particular place, we are aware of the extent of our own visibility. It is our place and space only, defined by our own interests, desires, and goals, as well as by our momentary feelings. The horizon encircles this place; no matter where we go and where we place ourselves in face of a landscape, the horizon constantly asserts its presence.

In a landscape, our location is determined by what we see, touch, and feel, and not by the totality that we are able to represent and to record as a system of positions, distances, and directions. We live in the present and enjoy or lament whatever the present offers us. Detached from abstractly represented locations, we may be lost – for good or ill – in a landscape. We know only what is near to us, what is offered to our senses, what elicits joy or fear. We orientate ourselves with the helpful support of trails, roads, bridges, rocks, streams, and ponds. In a landscape, we have to find our own way.

Szerb's account of his experience of the Italian scenery does not convey such a sharp separation between the concrete experience of a landscape and the abstract representation of a geographical environment. Unwilling to ignore the spatial and objective knowledge of surrounding regions, Szerb nevertheless understands himself as someone enclosed and embraced by the lovely mountains around him. His whole being is happily absorbed in a beautiful Italian landscape without completely losing sight of the inhuman

3 Straus, *The Primary World of Senses*, pp. 318-23.

and distant world and without forgetting that he is cheerful and complete only by leaving behind a hectic group of sightseers and their mundane interests.

The author of *The Third Tower* relates his joy of savouring "the inexpressible sweet serenity of the Italian landscape." A natural environment arouses in him a higher degree of sensory alertness than the ambient visual and auditory noise of the city he has just left behind. As he sits calmly at the foot of the tower, away from all the chattering tourists, he notices the animated presence of a landscape. To be sure, under different circumstances, the beholder might detect, in the surrounding mountains, entirely different qualities. He would perhaps speak of the "cold immensity" or "remote unfriendliness," or simply "melancholy" of the surrounding mountain range. A person who happens to see the Apennines covered with snow and to feel the icy wind gushing through its valleys may not echo Szerb's glowing view of the landscape. For instance, for Aldous Huxley, who attempted to travel from Florence to Bologna in an inclement season, the Italian scenery was a rather disconsolate and unfriendly sight.[4]

As with the material aspects of the landscape, this particular and dynamic quality of a landscape is subject to ongoing transformation. The same scenery may appear, in different seasons, as either warm, joyous, inviting, ominous, oppressing, or hostile. Whatever the detected emotional quality may be, it provides the natural surrounding with a certain intensity and an irradiating force. In every presence, we sense a certain power, which intervenes between the surroundings and ourselves and creates a singular reciprocity: we approach the landscape with a heightened awareness and the landscape, in turn, solicits an attentive sensory response from our part.

Each of these presences offers a different atmospheric impression. A snowy mountain on which skiers descend is entirely different from the hill enveloped by a dense forest. The atmosphere of feverish mobility is replaced by an ambiance of peaceful immobility. As we move through a meadow and remain alert for subtle and complex

4 Huxley, *Along the Road*, pp. 190-204,

perceptual cues, we detect the underlying affective quality of our specific surroundings. We might follow the example of Szerb and describe the atmosphere of the mountains as "human-like" or, while we walk through a pine forest, apply the concepts of mystery or starkness. The atmosphere of a landscape consists of a specific "emotional essence" diffused throughout the material elements. We instantly apprehend this essence, and each subsequent and explicit visit merely confirms the validity of our primary mode of perception.

A particular atmospheric quality is not identical with some objectively given traits. Yet we sense it as we come into a contact with a path, a tree, or a creek. It is everywhere and nowhere. Therefore, we are not always able to give a clear account of our ways of replying to situations. We may walk through a forest without being able to pin down why we feel ill at ease. We may refer here to a pertinent observation of Maurice Merleau-Ponty who related his experience of entering a living space (a city, a building, or an apartment) and immediately perceiving the character (*l'esprit*) of those who live there, without being capable of justifying this impression through any enumeration of notable details.[5] In a similar fashion, we detect the affective quality of a landscape: while, and even before, apprehending its physical features, we sense its incomparable and peculiar character.

But what is a human-like landscape? A landscape is endowed with a human character when we experience a correspondence, a suitability between the material characteristics of the landscape and our whole being. The fitness is made possible by the restraint which is exercised by our approach to the landscape as well in the manner with which the landscape presents itself. Conversely, we may speak of the absence of connectedness when we notice the landscape's colossal size, overpowering forcefulness, and crushing presence, and, indeed, its lack of restraint.

A landscape displays its restrained presence when it gradually opens up, lifts its veil, without revealing all its facets. While sensing this restraint, we repeatedly return to a park or to a heath

5 Merleau-Ponty, *The Structure of Behavior*, p. 173.

and happily discover what we have missed before. In another sense, the landscape's restraint is apparent when an awe-inspiring and vast natural reality – a forest or a mountain or a large lake – does not generate distress and terror. We have the impression of being in the presence of a potentially frightening but tamed power. It may happen that we stand on the top of the mountain and feel the presence of a considerable amount of energy below us. Yet we know or at least hope that this energy does not constitute any menace or threat to our existence.

The human-like landscape solicits the restraint of the beholder. We impose a restraint on ourselves when we perceive a sensual harmony or ecological balance of the meadow and we want to preserve it and respect its peace and stillness. Obviously, an alteration of the landscape may occur in due course. However, such a change may be introduced with a sense of discretion, without destroying or altering the harmony of the whole. Then we are careful about what we add to the landscape as well as about what we would like to take away. In a word, we hold back a power we have; we do not use it. This human capacity is captured by the French term *litote*. *Litote* is the ability to hold a power in check and exercise restraint; it is the guarantor of refinement, elegance, and balance in whatever we decide to touch and transform.[6]

We may consider the silent presence of a landscape as a form of restraint. But this kind of silence does not entail the absence of sounds. The steeply descending brook and the gently ruffling leaves on the trees produce sounds. Beyond these sounds, we still perceive the silent atmosphere embracing everything in the forest. Even if Huxley hears "fearful noises" as he approached the town of Pietramala, the slopes near him exude calm and stillness. Silence is contrary to whatever is obtrusive, rowdy, petty, raucous, and wicked. Unruly and boisterous people, leaving behind them a mess, clash with the silence of a landscape. The gestures and movements of an agitated person setting up a camp in the open air can be noisy and unruly even if he does not emit a single sound.

6 Serres, *Le Tiers-Instruit*, pp. 177-93.

In his essay on the world of silence, Max Picard contends that the silence of nature is a "primary reality."[7] The segments of a landscape make the original silence "clearly visible." We tend to associate silence with darkness and the daylight with the sounds of conversation or of music or of transport vehicles. Yet, in the heat of the summer noon, silence absorbs everything in a landscape and dissolves itself into light. Then, "the light seems so much the essence of silence that the word seems quite unnecessary. The light is all at once the fulfilment of the silence."[8] As time advances, the silence of the night lends breadth, mystery, and splendour to the trees and the starry sky above them.

In his novel *Butcher's Crossing*, John Williams contrasts the silence of a pristine valley with the noisy and wicked slaughter of buffalos. In fact, he juxtaposes goodness and bestiality, humanity and inhumanity. The landscape's dense stillness portends the intensity of the cruel destruction that the guns of the hunters will soon achieve. The image of these men standing quietly above a beautiful valley makes, for the reader, the eventual senseless and unrestrained killings and the correlate profanation of the landscape more brutal and more devastating.

> Lush grass grew on the bed of the valley, and waved gently in the breeze as far as the eye could see. A quietness seemed to rise from the valley; it was the quietness, the stillness, the absolute calm of a land where no human foot had touched. Andrews found that despite the exhaustion he was holding his breath; he expelled the air from his lungs as gently as he could, so as not to disturb the silence.[9]

Although people now rarely find themselves in the presence of a virginal landscape, this type of experience is far from being uncommon. Genuine hikers and passionate excursionists are sensitive to the silent presence of a distant mountain or the encompassing calmness of a green meadow. They are eager to preserve this primary

---

7 Picard, *The World of Silence*, p. 137.
8 Ibid., p. 141.
9 Williams, *Butcher's Crossing*, p. 117.

silent presence of the landscape while they reach the summit of a mountain and admire the valley and the surrounding peaks.

Szerb relates his sense of well-being and wholeness when he detects perfection in the contemplated scene. He declares that all is well, nothing is missing; his existence is complete, which brings him the feeling of gladness at being alive. All the tensions and worries of his everyday life have been for a while forgotten as he enjoys a euphoric moment of harmony and fulfilment. The feeling of well-being are lived as a grace; it occurs without calculation or control; it eludes the power of the individual to produce it at will; it induces cheerfulness and gratefulness; it creates the impression of beauty stirred by what the beholder sees as a perfect landscape; all its elements appear vivacious and lively.

A landscape is seen as a perfect presence when its elements and proportions are congruous with our feelings of peace and elation. These feelings tend to be absent when we are confronted with the immensity, austerity, excessive dimensions of a mountain or of a forest. To judge a landscape perfect, we have to view it as an animated reality, to which we somehow belong and in which, occasionally, we are positioned with ease.

There are lovers of undulating hills, lovers of peaceful meadows, and lovers of the sea's "deep and calm solemnity" (Søren Kierkegaard). As John Cowper Powys remarked, "There is undoubtedly a deep affinity, probably both psychic and chemical, between every individual human being and some particular type of landscape."[10] As we have seen, Szerb experienced joyfully this affinity with the Apennines and qualified it as human-like while attributing to it the alteration of his whole being. He extends this quality to the entire Italian landscape and considers it "more beautiful" than any other landscape of the world.[11]

How should we understand the often-repeated and spontaneously voiced claim about the beauty of a landscape? Should we speak of

---

10 Powys, *The Meaning of Culture*, p. 150.

11 I have translated the Hungarian *szebb* with the English "more beautiful" and not with "lovelier," which is proposed by the translator of the book.

beauty only when the beholder is able to view a scenery with a sense of aesthetic objectivity? In other words, should we make the presence of beauty dependent on the adoption of a distancing, objectifying attitude towards a landscape? Is the intensity of our perception of the beauty of a scene also heightened by our sense of contrast with other scenes? Is the silence of the mountain is strengthened because we normally spend so much time in an urban jungle of concrete?

We find illuminating first-hand accounts of the experience of beauty in presence of landscapes in Timothy Beardsworth's book.[12] One of the contributors, who sat opposite to "a beautiful tree covered with pink blossom, thrown into a relief by the high hill behind it," speaks of being "caught up into the reality and being "part of it." Another individual was "in presence of a great natural beauty" and felt that he "was extending into (his) surroundings and was becoming one with it." He had a "strange feeling of expansion," an experience "difficult to describe." Again, another individual spoke of being "caught up into a tree," becoming "part of it." "He "was extending into his surroundings."

The contributors identify and praise beauty less by following an aesthetic canon and considering the formal facets of trees, meadows, or hills. They speak of an altered relation to the grass, the tree, an orchard. The beauty of the landscape is found in the experience of synchrony with the surroundings and in the significance that a person spontaneously attributes to a scenery.

Beautiful is a grove, which prompts a sense of unity, induces a wonderful feeling of freedom, makes one simply stand in muted awe while dwelling upon a scene that prompts the full enjoyment of the moment. Beautiful are the leaves, the field, or the surface of the river because they are alive, mysterious, "infinitely significant;" they bring the person to "the verge of a great revelation." One contributor stated: "I felt if I held my breath I would hear breathing." Whereas the perception of formal qualities of order, balance, complexity, proportion, clarity, rhythm merely lifts one's spirit, the "memorable

12 Beardsworth, *A Sense of Presence*, pp. 48-64.

experiences of one's life" create an intensification of the person's sense of being absorbed by a tree, a meadow or a creek.

All these accounts mention an unusual sense of peace and well-being, a previously unknown surge of unspeakable joy, an acceptance of everything with a "sort of glorious indifference," an "inspiration for creative activity," "sheer exhilaration," an "attentive, breathless synthesis" between two realities as if the "I and what the I was watching were one."

If there is a condition necessary to live such an experience of presence in the face of a landscape, it does not reside in the adoption of an attitude of aesthetic objectivity. A prerequisite of the perception of a beautiful and perfect presence is rather the distance taken from other human beings. One has to absorb the impressions alone, in a state of full composure. Szerb asserts that by leaving behind his tourists companions, he is able to come into possession of his soul, to be in touch with his own centre, and recognize the Italian landscape in its human-like beauty. Contributors to Beardsworth's volume also remark that a close personal contact with fellow human beings "destroys" the ability to relate to oneself and the surroundings with the sense of unity and fully enjoy the encounter with the intensity of beauty.

Can we enjoy a similar experience when we come into a contact with a city? How does the presence of a city assert itself?

Here is an experience of presence taken from Beardsworth's enquiry:

> I was walking along a road in my home town (Eastbourne) approaching an incline where the road branched and straight ahead was an old flint wall covered with ivy. I had the feeling of everything opening out (or a veil being lifted) and a wonderful feeling of freedom (lightness) and of comprehension of everything. It is so hard to describe.[13]

Taking a morning walk on the streets while the shops wake up the community, standing in the middle of a square at midday and looking around at the buzzing traffic, finding a peaceful moment of

13 Ibid., p. 54.

rest on a playground, in an outdoor garden or in a cemetery, entering and leaving office or residential buildings, visiting places of worship and observing all their architectural intricacies – all these moments provide us, in succession, with a particular experience of the city. Each of these illuminating perspectives makes us apprehend the same general style, flavour, emotional essence, the city's distinct atmosphere. As we walk and take pleasure in the direct experience of a particular place, as we meet all sorts people and glance at their work or leisure occupations, we realize that each change in perception gives us not only a different aspect of the city but also the successive confirmations of the city's unique emotional tone. Wherever we happen to be, our bodily sensibility detects in the city an all-pervading sadness, nervousness, energy, calmness, cheerfulness, or serenity and we feel deeply affected by a distinct affective quality. It is not always easy to articulate with concepts this invisible "soul" of the city. Nevertheless, each street, each house, and each walking staircase bring the indelible validation of our initial affective experience of the city's singular character. Each of these places may yield to an experience of presence.[14]

We may encounter the presence of a city by looking out of our window or by stepping out of our car at a gas station or by strolling on the street or by climbing up to a rooftop. However, let us now study this experience of presence from another point of view: gazing at the city from a bridge on which we stand.

Many cities and towns have been built around a river or around the junction of two rivers. With time, the increase in the number of houses and their inhabitants' movements have given rise to the need to erect bridges. It may happen to us that during a warm and sunny day, we decide to go over a bridge and from this particular place to peruse a city disinterestedly. Instead of crossing the bridge and scurrying to a specific destination, we just stop walking and stand in the middle of the bridge besides the railing.

Here, as we interrupt our stroll, the bridge receives an entirely new significance: it is no longer seen and used as a means to go

---

14 Ladrière, "La ville, inducteur existentiel," pp. 144-6.

somewhere, to overcome the separation and distance between two points. Its overall worth is no longer defined by the criteria of usefulness. The bridge presents itself with its intrinsic meaning and not with its otherwise important practical purpose. As we make a voluntary halt, we apprehend the bridge's aesthetic and functional values. We see, from above, the pillars and the statues placed on pillars, the span by looking sideways, and the railing by leaning on them. Our vision does not glide over their design but comes to dwell on it and to grasp its constitutive aspects distinctly. I often marvelled at the slightly Y-shaped form, at the broken axis of the Margaret Bridge in Budapest, one of the few bridges having such a construction design in the world. The direction, thickness, and frequency of the pillars and the corresponding proportion of the span length and bending have been adjusted to the power of the Danube flowing from both sides of the Margaret Island.

Henry Wadsworth Longfellow's poem *The Bridge* brings a reflective tone to the human desire to stand on a bridge. Two stanzas taken from the poem bring forth some of the features susceptible of conveying an experience of presence.

> I stood on the bridge at midnight,
> As the clocks were striking the hour,
> And the moon rose o'er the city,
> Behind the dark church-tower.
>
> How often, O, how often,
> In the days that had gone by,
> I had stood on that bridge at midnight
> And gazed on that wave and sky!

As the poet stands on the bridge, "the thought of other years" come to him. Standing on the bridge after a weary day, during the night, allows him to reflect on how the "sorrow of others" touched him, how many persons, young and old, "hot and restless," "subdued and slow," crossed the bridge in the past, and why grief and hope fill the heart of all those men and women who cross over the bridge back and forth. But the bridge may serve another function: it stops the forward moving person and stifles the purposive intention of focusing on the future; it invites the poet to gaze at the familiar realities of his life in the present: the moon in the sky, the dark church-tower, and the flow of the river.

Another person, standing on the bridge at midnight, might be seized by the silence reigning over the city. There are no citizens walking on its sidewalks and there are no vehicles speeding on its streets. The buildings, squares, and arteries are enveloped by a melancholic dispassion. The throbs of the city came to a full stop. Yet the flowing river and what it reflects express the presence of life, the never dying passion of young and old, their woe and love.

> And forever and forever,
>   As long as the river flows,
> As long as the heart has passions,
>   As long as life has woes;
>
> The moon and its broken reflection
>   And its shadows shall appear,
> As the symbol of love in heaven,
>   And its wavering image here.

In a city, the bridge brings *openness* to our vision. While the buildings along a street either discourage looking above the horizontal or hide a large part of the sky, the bridge is the place where a much greater part of what is elevated can be contemplated. By standing on the bridge, we are able to see, from a distance, the expanse of the clouded or luminous sky with all its colours: the intensity of the blue is mixed with the play of purple, orange, pink, yellow, and red. Our impression of the large sky keeps changing since the appearance of a dominant colour and the combination of other colours depend on the sun's actual position.

Looking at the boundless sky may evoke a sense of vastness, which contrasts with the perception of the narrow confines of the bridge and of the natural boundaries of the river. The sky suggests the free realization of all our aspirations, while the bridge makes us keenly aware of the limits of our human condition. The sky upon which we direct our gaze and the bridge on which we stand firm bring to mind the ongoing tension between our most personal impulses and the obligation of exercising restraint on them. The sight of the open sky further captivates us when, ahead of us, our attention is drawn to the sun as it rises or sets. The former conveys morning's energies, the latter twilight's serene resignation. The wonder solicited by the slow emergence of daybreak or the coming of nightfall prompts in us the curious impression that time both comes to a stop, anchoring us in the timeless present, while at the same time advancing indubitably, pushing us forward into the future.

The bridge is above something: either the land or the water. The quality of *aboveness* is a pivotal aspect of the bridge. The bridge elevates us, even if this elevation seems, with certain structures, hardly significant: it puts us above the river, the lake, the ocean, the creek, the road, the rail, the field. The essential aboveness of the bridge means that we are able to gaze down at the river below. We may stare with admiration at the presence of the calm Rhine in Cologne or of the torrential Sora in Škofja Loka, Slovenia.

By moving downstream softly or ferociously and staying in its banks, the river reveals a singular tamed power, which reminds us of the restraint of the landscape. Standing on the bridge, we see below how swiftly the river's current takes large wooden branches or trunks downstream. We notice how much speed it adds to a moving boat. But a river following its natural channel, carrying away objects, giving pace to the motion of a boat manifests another kind of power. Perhaps more convincingly than the ticking of a clock, the mildly or mightily moving water brings to the fore the power of time, which flows like the river, advances ineluctably, and leaves behind everything experienced in the present.

In his famous poem *The Mirabeau Bridge*, Guillaume Appolinaire introduces the metaphor of the flowing river when he bewails the passage of time and the fading away of love. But the poem's refrain also voices forcefully the claim: I remain present. The poet asserts his presence in the face of all the upheavals, joys, and sorrows that life plays on him. The reader might picture him being present on a specific location: on the bridge or on the bank beside the river. The poem hints at the significance of the bridge above and the river beneath: while the former stands firm, undisturbed and holds the poet in the present, the later seeks the change and hastens the present to become the past.

Beyond the sky above and the river below, we may become attentive to the presence of the city ahead of us of as it spreads on both sides of the river. We are then taken by the vision of a harmonious whole made up of public buildings, churches, cafés, broad avenues with trees, moving vehicles, tramways, cars, bicycles, the curves of the river, docks and boats. Men and women walking speedily or just standing or sitting and sunning themselves or watching the flight of seagulls. It is not unusual to see a city, from a bridge, with contrasting features: a picturesque range of hills on one side, flat plain vistas on the other; areas covered with solid and sombre stone buildings alternate with lively green parks, filled with paths, trees, flowers, birds, and running and laughing children under the watchful eyes of their parents; historic churches adjoin modern structures, a hotel, a school, or a theatre.

We may come to a singular experience of presence, when we approach the city not as a restless tourist who sees everything through the lenses of a mobile phone or a weary inhabitant who notices only the approaching vehicles, but as a poet or a painter who glances leisurely at whatever meets his eyes: playing children, blossoming trees, birds chirping on branches, or waves hitting the banks. Then the city suddenly animates as a living form, similar to the form of a painting. It presents a meaningful, dynamic, and balanced structure with rhythms, shapes, colours, sounds, contours, pauses, reliefs, outlines, patterns, and shades. The consciously built pictorial whole is filled with vivacity and animation; it receives an entirely new significance. We have the impression that the river, the houses along the banks, and the sky come fully alive; they are luminous and irradiate an active and intense energy. We apprehend all these things in their completeness together with the manifold connections they entertain with each other and with their proper environment. We no longer perceive isolated and selected realities such as the movement of a boat or the faint sound of a tramway. We deal with the whole, the complex interaction of meaningful and lively forms. We find a pleasure in the presence of a picture-like interaction of forms – the wealth of shapes, figures, and tones as they complement and contrast each other, the rhythm of lines and designs, the interplay of lights and shades, colours of all sorts.

In a beautiful poem, entitled *Composed upon Westminster Bridge*, William Wordsworth tells us how he marvelled at such a magnificent presence:

> Earth has not any thing to show more fair:
> Dull would he be of soul who could pass by
> A sight so touching in its majesty:
> This City now doth, like a garment, wear
>
> The beauty of the morning; silent, bare,
> Ships, towers, domes, theatres, and temples lie
> Open unto the fields, and to the sky;
> All bright and glittering in the smokeless air.
>
> Never did sun more beautifully steep
> In his first splendour, valley, rock, or hill;
> Ne'er saw I, never felt, a calm so deep!
>
> The river glideth at his own sweet will:
> Dear God! the very houses seem asleep;
> And all that mighty heart is lying still!

The poet felt the pleasure of gazing at the presence of a city when, early in the morning, perhaps as he was standing alone on the Westminster Bridge, he set his eyes on the magnificent sky, the animated river, and the quiet, calm, and still sleeping houses. He looked at these realities not as a mere resident who leaves his flat to fetch something but as a poet who contemplates a living form, displaying complementary human-made and natural elements, with an artistic frame of mind. He brought disparate elements of the natural and urban landscape together and, in an extraordinary sonnet, related his wonder at the sight of a city. His instantaneous vision of his beloved city induced in him a profound sense of well-being and peace. Like Antal Szerb, he felt a tremendous stirring within himself. I fancy, it is fair to say, that, for them, these exquisite moments made life worth living.

Things

Many years ago, during a serene summer afternoon, I was sitting on the grass, under a tree, beside an outdoor swimming pool at the Aquatic and Recreation Centre in the German city of Krefeld. Close to me, a young boy, of age of twelve or perhaps a little older, was eating his sandwich. When he finished his meal, he took a glass bottle out of his bag and opened the bottle's metal cap with a small opener. He put down the bottle opener beside him on the grass. I glanced curiously at the thing. A rather small, round handle was attached to a rectangular part. I was curious about the handsome device and asked the boy if I could have a closer look at the handle. It was an oversized replica of one *Pfennig*, the smallest German currency coin of the time. After I have expressed my pleasure at seeing such a well-crafted tool, the boy lifted up his head, looked at me and said: "Do you like it? Here, I'll give it to you gladly." Following a few moments of hesitation on my part and much insistence on his part, I accepted the bottle opener and expressed my sincere gratitude for the generous gesture.

I still use and cherish my one *Pfening* bottle opener. Made of solid copper, it has suffered no damage over the past numerous years. Holding it in my hand and feeling its smooth and round sphere evokes a particular satisfaction, similar to the contentment elicited by a ball or any other round-like object. The ease and speed with which I open bottle caps further enhances my pleasure.

Now, however, when using it, I occasionally find myself pausing, delaying my movement, and gazing upon the device. It, then, yields an experience of presence. I notice vividly what the thing *is* in itself regardless of its purpose and of how and when I acquired it. I see neither its useful feature nor the generosity of the person who gave it to me. I see it as an object that irradiates its being, it becomes brighter, it presses itself upon me with its form and material and excludes everything else from my vision. I am filled with a sense of peace and of alertness as I hold the device on my hand. Everything around me is all right. A minute or two later, however, I see the kindness, the generosity, the sense of disinterestedness of a young person, who was ready to give up something valuable in order to bring pleasure into someone's else life. I see the heart-warming present, which was given to me, by an unknown person, graciously and without any desire of receiving something in return.[1]

A genuine gift is a sign conveying an interpersonal attitude and this attitude becomes manifest in the act of giving. It may be a sign of thankfulness, recognition, kindness, or the simply desire to bring pleasure into someone's life. This attitude, altogether different from material characteristics and modes of use, brings to the thing a new dimension. It puts it under a different light; it creates radiance; it makes it speak, to be sure, without words.

In one of his luminous essays, Eugène Minkowski has shown that, in certain circumstances, a thing may exhibit an aura, a breath of life, a soul. It is no longer a thing that we consider uniquely in terms of daily consumption, or its pragmatic value; we no longer put it easily aside when its use is no longer required, or throw it away if it becomes useless. It integrates itself into our "spiritual becoming," exerts an impact on it. It may be seen with characteristics entirely different from what we habitually deem important in our daily life.

> A thing may become animated in front our eyes, and glow, vibrate with new life, entirely different from all the habitually recognized particularities, thus emerging, without any palpable transition, from life itself and exhibiting, as a coat woven with hardly perceptible threads, all its poetry.[2]

---

1 Denis Grozdanovitch had a somehow similar experience during his stay in Québec City. *La gloire des petites choses*, pp. 13-15.

2 Minkowski, *Vers une cosmologie*, pp. 248-9.

In these occasions, as we see the world around us, as it were from a certain distance, the thing may yield an experience of presence. It appears with all its luminosity, vitality, and with a particular emotional tonality. The thing becomes a "bearer of something" that does not usually come forth. We have then the impression that it carries a depth, a certain spiritual force, a life, a compelling atmosphere.

Such transformation occurs in the lives of children. In their play, they give to things a new significance: a chair becomes a rowboat, a broom an oar, a large piece of cardboard a rudder. Children play with them because they see them as animated realities, appealing "images," presenting sensory and motor possibilities while communicating suggestions or demands necessitating that they are seen and handled in a certain way. They are things which are worthy of their outmost care and genuine love. One needs only observe the gentle handling with which a child moves a doll into her bed. Their play unfolds thanks to the reciprocal interaction between the chair-boat and the child. Children, then, appear to become wizards who animate their objects of play through their behaviour, approach, movement, or word; things are endowed with a life of their own, with a singular presence, exerting on them a fascinating influence and inviting them to pursue their enjoyable play with the utmost ease, in a timeless present. In absence of this living presence of playthings, no play can occur and unfold.[3]

I take a book from the shelf: It is a solidly bound hardcover volume. I enjoy holding it in my hand a durable thing that I have kept for many years. I examine the image on the cover; I look at the pages, I see which typeface was used, how the lines and of margins are disposed; I smell the odour of the paper; I sense its texture as I turn the leaves. The pleasurable sensory experience of the physical aspect of the book may invite me to recall the moment of its purchase and the store in which I have found it. I noted down the time and place of its purchase on the first page. I see myself entering the store, picking up books one after another, reading their table of contents or index, and leaving the store with the chosen one in my hand. I think

3 Buytendijk, "Das menschliche Spielen," pp. 93-6.

of the care with which I preserved it over the years and the joy I felt when I sat down and read it. It evokes the thoughts and emotions I was able to draw out of its pages and the moments in time which allowed me to share these thoughts and emotions with my students.

Another book brings into my present the author who gave it to me as a gift. I find in it the presence of the author as I read now the affectionate words of dedication on the flyleaf. The words and the generous gesture of offering me a present animate and enlighten all the pages.

When I hold the book, I disregard, for a moment, the thing in my hands and focus on the experience that I lived in the past. Through my act of remembering, making an effort of reliving fully the experience of receiving the gift, I make the past to be present. My visualization of a past happening, which I consider enjoyable, can entirely hold my attention and therefore, fill completely my present. As I put the events of my past together, the fact that I experience the presence of the book here and now does not hold my attention. The past with all its emotional content becomes a captivating presence. This visualized past may be summoned at will or may surface spontaneously by itself. By singling out a precious and dear moment of getting a present, it may push aside all the other experiences, perhaps all the griefs and disappointments tied to the past.

It may happen that the past to which I return is a succession of experiences in relation to the same thing. I may recall how a thing appears on different days, how its significance changes with time, and how it displays its presence with a variety of intensity and value. The lottery ticket appears different when I purchase it and when I learn that I have failed to win a large amount of money. I see the same thing in two different contexts and in connection with two different emotional states. For a historian of the Victorian era, a lamp-post in Paris is more vivid and more alive than the lamp-post in London.[4] The visualized presence of the thing cannot be considered detached from a particular significance tied to the relation of the thing to a particular environment.

---

4 Chesterton, "Lamp-Posts," pp. 10-11.

A poem by the Hungarian poet Lajos Áprily, *Milk-loaf, Tuesday milk-loaf*, recounts the extraordinary presence of a sweet and fluffy bread as it appears in retrospect on different successive days. The light it radiates and the scent it emits creates a festive, bright, and carefree atmosphere, which permeates the whole Sunday and even the days beyond. As the loaf is gradually eaten, the precious food changes in appearance. It becomes a presence. On Tuesday, the mother kept hidden a piece of the loaf, but only for the child. Its singular presence is able to offer an intimate glow, which is powerful enough to keep alive the joyful atmosphere and to hold all sorrows, worries, troubles of routine life at bay. From a distance, the loaf is not only seen as a modest and tasty food, but also as a spiritual nourishment, making of average days sacred and bringing, in time of need, solace and consolation.

Let us look at the experience of a thing, an old medallion or a handsome pen, in the present moment. I can be receptive to the thing as it is something present in itself; it shows itself in its complete and determined self-sufficiency and not merely by reason of the force of my emotional attachment that I have developed over the years.

I am able to see the thing as it is *present in its own being* and, moments later, as being *carrier of a past emotional experience or a quality or a world*. It appears first as a rather distant and independent reality and, later, as a valuable, intimate, and familiar possession. I realize that the thing may impose itself with its being. It might prompt me to ask the question: why it is there now, in the present, with its intensity, its particular colour, its sharper outlines, and its singular form? Usually we do not stop and wonder about the fact that a thing is. We take its being for granted. When I gaze at the actual presence of the thing and, with a sense of wonder, see it for what it is, it appears as a tangible, solid, concrete, and stable reality, the emotional aspect of the thing does not disappear. It is merely pushed, for a while, to the background of my awareness. Since I focus on its being and not on a certain past experience, the thing allows me to experience the pure present. "Experiencing the present purely," writes Annie Dillard, "is being emptied and hollow; you catch grace as a man fills his cup under waterfall."[5]

In what other forms does the presence of thing appears to us? What are the qualities of things that an experience of presence brings forth?

We are surrounded by an endless variety of things. As we move from our home to our workplace, from our workplace to our leisure milieu, and further to other environments, we enter into contact with chairs, tables, cups, plates, household devices, vehicles, doors, handles, windows, benches, bicycles, and many more things. Usually, when we feel well, we focus on our immediate task and pay little attention to the things we need to handle or use. We use them almost unconsciously by summoning up the appropriate habitual movement. Likewise, our hands or legs, performing a great variety of movements with energy, practical efficiency, coherence, and skill, also remain unnoticed. We notice the thing if it does not fulfil its purpose or our body if it fails to carry out the assigned task.

But when we look expressly at things not only in themselves, but in direct connection with our active body, we see how they are made for being handled, grasped, pulled, pushed, lifted, lowered, pressed by the body. Their design and form are suitable for an eventual bodily action. As we observe a cup, we remark its availability for its immediate and comfortable grip by our fingers. Likewise, the already mentioned bottle opener offers itself for an easy action of pressing and lifting. Things assert their presence with their *quality of fittingness*; their form highlights their suitability for bodily action. Fittingness means, in this context, the correspondence between the form of a thing and the diverse movement possibilities and the form of our own body. Sometimes we notice how well our fingers are able to hold a pair of scissors while cutting a piece of paper or how effortlessly our hand presses the curve of the handle in order to open a door and enter into a house. The versatility of the fingers, the skilful gesture and the thing to which the gesture is keenly adapted provide

5 Dillard, *Pilgrim at Tinker Creek*, p. 82.

us not only a pleasant sensation but also an occasion to marvel at the presence of realities fitting so well one another.

There are things in our immediate surroundings, which present themselves for multiple kinds of functional use. A brick may be placed above another brick to build a wall or combined with other materials to create an artwork. The brick then looks different for a mason and an artist. And since, in the eyes of the user, a thing takes a different function and a different shape, their presence is obviously different. For both, the thing is not merely there in its own being but also together with its instrumental or aesthetic value.

Since the thing's presence is underscored by the quality of fittingness, by how the thing and the body adjust to each other, the body also takes a different form. Its form is in harmony with the task at hand and the type of movements that leads to a satisfactory achievement of the task. It takes a different form as it leaves behind one kind of thing and turns to another. The change usually occurs spontaneously, without special attention or voluntary choice. We may say that the legs and arms and other parts of the body are forming themselves. Each time, while executing a particular form of movement in relation to a particular thing, the figure of the body modifies itself, it forms itself. It takes on a certain form. We may think here of the difference between tending flowers in a garden, playing a musical piece on the piano, and jumping after sliding down on skis on a curved ramp. We then rightly speak of the gardening body, the music making body, and the ski jumping body. Although each person approaches a particular task in an individual manner, the forming of the body happens according to a model indicating how to accomplish tasks prevalent in a natural, artistic, or athletic milieu.[6]

The previous examples made clear that the form and skill of the body are, in fact, also presences, which go together with the presence of things. In order to experience the fitting presence of a thing, we have to be present to ourselves. However, this self-presence is usually not in the forefront of our awareness. It does not arise as an object of self-observation. The presence of the skilfully moving

---

6 Berg, *A Different Existence*, p. 58.

body is merely felt during the execution of movements: the skill and form of the body is sensed as right and appropriate in a given situation. It may happen that the timing and result of the performance are particularly felicitous. Then, together with the achievement of the task, the skill and form of the body become an explicit presence.

In a *Paris Review* interview, Thornton Wilder defines his own literary task in these terms: "So that I see myself making an effort to find dignity in the trivial, against those preposterous stretches which seem to rob it of any such dignity, and the validity of each individual emotion."[7] We usually consider a thing trivial, elementary, or trifling when it falls outside the focus of our daily life. It is viewed as slight, negligible, something not worthy of our vigilant care and conscientious attachment. We may ignore it or pass over it and still achieve what we consider are the central tasks and activities of the day. We may converse with a neighbour about a trifle issue or take part in a trivial event. The issue and event leave no lasting effect on us. They lack relevance, weight, and import. We tend to ignore a trivial concern, belittle a small and trivial accomplishment, or look down upon a trifle interest in something.

Still, even if treated with disinterest or even contempt, what makes the presence of the trivial aspects of our life, under certain circumstances, eminently noticeable and stirring? How and when do we remark the glowing presence of a thing perceived with the *quality of triviality*? What makes a trifle, a pebble or a piece of paper, so dear to us?

In Wilder's play, *Our Town*, one of the main characters, Emily Web dies during childbirth and joins the dead souls in the cemetery. Against the advice of the others sitting beside her and with the help of the stage manager, Emily goes back among the living persons to relive a happy day of her life, the day of her twelfth birthday. As the morning encounter with her parents takes place, she sees everything with different eyes. Witnessing the kindness of her parents and the charm of the most ordinary things, she finds the experience too painful and asks the stage manager to take her back to the cemetery.

7 Cowley, ed., *Writers at Work*, p. 114.

In this moving scene, Wilder tells us that we often fail to see the smallest events and trifle things in our daily life. We close our eyes to what is the closest to us, what has become so familiar such as, in Emily's life, clocks ticking, sunflowers, food and coffee, ironed dresses, hot baths, sleeping and waking up. "I didn't realize," says Emily, "So all that was going on and we never noticed."[8] Wilder suggests that we perceive and remember the presence of trivial things when we consider them from a distance, when we approach them outside the perspective of conventional life, when we experience them in the painful state of desynchrony. Emily knows that she does not belong to this particular world any longer and yet she is allowed to recall things from the point of view of a guest. Things are so forcefully appealing to her: "I can't look at everything hard enough."[9] Although the food and the dresses have no function in her life anymore, they communicate their magical presence and, in fact, they tell her what was the most important in her life: the love and care expressed through the cherished things and delightful practices of everyday life. It is this previously unnoticed and now recognized presence and significance, which fill her heart with unbearable pain.

Vladimir Nabokov was the most appreciative poet of the trifling matters that I know of. Here is one of the many instances where he discusses the significance of trifles in the lives of poets and children:

> What exactly do these irrational standards mean? They mean the supremacy of the detail over the general, of the part that is more alive than the whole, of the little thing which a man observes and greets with a friendly nod of the spirit while the crowd around him is being driven by some common impulse to some common goal. [...] This capacity to wonder at trifles – no matter the imminent peril – these asides of the spirit, these footnotes in the volume of life are the highest form of consciousness, and it is in this childishly speculative state of mind, so different from common-sense and its logic, that we know the world to be good.[10]

8 Wilder, *Our Town*, p. 100.
9 Ibid., p. 97.
10 Nabokov, "The Creative Writer," pp. 193-4.

Illness creates a similar relation to things: it pushes things outside the intimate and familiar horizon of existence and, by doing so, it turns the gaze upon them.[11] When I am ill, I notice all the things I like to surround myself. They are fully present and I see them more clearly now than ever before. Illness puts them at a distance as if they were no longer part of my life. As I leave my bed for a short time, I see the table covered with various things: a lamp, a telephone, a computer, pens, books, papers. To my great regret, I am no longer able to make use of them; I must return to my bed, rest and sleep, and get well again. Things on the table fall outside of my current sphere of interest. They are part of my healthy and busy existence, which is now further away from me. The presence of things appearing with their *quality of elusiveness* reminds me of what I have taken for granted, my health and all the things I was able to use in my state of well-being. I see them now as if they belonged to someone else who, eventually, will carry out the work with their indelible support.

There are well-crafted and sturdy things, a tool or a piece of furniture, which remain around us for long time. Occasionally we pause and relish the reliable service they provide or the lasting bond they form with us. But, equally, we may show a similar strong attachment to things, which are fragile and prone to fading away. J.H. van den Berg believes that large and enduring things impress us but fail to evoke an unwavering emotional connection:

> What really touches us are the little, the futile, the very perishable things. Man is a creature of trifles; things that slip between his fingers are valuable to him, they will stay with him. What is perishable appears to enfold everything. Things are valuable to us–our things, things of our existence–because they are caught in a fast and radical deterioration. They need our nearness, and so they are near to us.[12]

In this context, we may think of a piece of bread, a feather, a figure traced in the snow, or even a water bubble. Is their nearness, their presence valuable for any other reason than their ineluctable and rapid

11 Berg, "The Meaning of Being-Ill," pp. 385-6.
12 Berg, *The Changing Nature of Man*, p. 94.

deterioration? Through our contact with these things, we become more intensely aware of the power of time, which is, in a way, the power of destruction, of deterioration. From the perspective of time, certain things reveal their inherent fragility, evanescence, finiteness. Their destiny is to be and, after a while, to dwindle, slip away, become eclipsed. Their time is up; the force of time pushes them to their waning. Because they are in the shadow of vanishing, we are inclined to cling to the presence of things having the *quality of being ephemeral*. We stare at our footprints in the sand until the incoming tides wash them away. We observe the motion of the oak twig that the autumn wind throws in front of us and, after a few moments, carries it away.

Nevertheless, we do not dwell merely on things, which fade away with the flow of time. Our eyes may pause on things, which seem to resist nature's destructive forces. A rake, a spade, a wheelbarrow, standing in the corner of a shed, appear a little weathered, but also with all their protracted duration. Dazzled by the bright light coming from the open door, they seem to be shielded from the wear and tear of time. Taken up by their presence and finding ourselves in an intimate communion with them, we step out of time and linger in the immobile now. Time, with its impetuous and irresistible flight, has no longer power over us. Everything is calm and silent around us and the presence of tools having the *quality of ever durability* put us in the timeless present.

Francis Jammes beautifully expressed this intimate communion with durable things in his poem *La Salle à manger* (The Dining Room).

In Janine Canan's translation:

> There is a rather dull cupboard here
> that knew the voice of my great aunts,
> that knew the voice of my grandfather dear,
> that knew the voice of my father too;
> and to these memories it is true.
> You're wrong to think it can only sit,
> because I talk with it.
>
> There's also a cuckoo made of wood.
> I don't know why its voice is no longer good.

I don't like to ask… You see,
the voice might really be
broken, up there in its head,
just like the voice of the dead.

There's also an old buffet
that smells of wax and of conserve,
meat, and bread, and ripened pears.
It's a faithful servant and it swears
'twill all our goods preserve.

Lots of men and women have been here, near it,
who do not believe in such a spirit.
And I smile, when a visitor comes my way,
that he thinks me alone, and starts to say:
"Good morning, Mr. Jammes, how are you today?"

Things are around us with their form, function, and significance. They also unveil a *world* if we dwell actively on their presence. The world comprises things, persons, institutions, stores, towns, cities, villages, landscapes, private and public spheres, modes of living and moving around, attitudes, practices, norms, conventions, laws, and all the complex and intricate relations among these realities. Whether we notice it or not, a whole world is constituted in relation to the thing which arrests our attention. This world can be familiar or foreign, friendly or intimidating, well known or unknown. When we encounter a thing belonging to a foreign world, elements of our own familiar world remain present as a horizon in the background against which we perceive, compare, and evaluate whatever we identify in the foreground. This comparison and evaluation are possible since the foreign always shows a singular degree of similarity with the familiar.

During his visit to Japan, while observing men and women eating their food, the philosopher and semiologist Roland Barthes became attentive to the presence of a common thing: the chopstick. It prompted him to compare two ways of consuming two sorts of food: bringing the small portion of Oriental food from the plate to the mouth by using chopsticks and eating copious Occidental meals by cutting and lifting the food by forks and knife. Designating, gently dividing and separating, carefully raising and carrying small amount of food are among the chief functions of the chopsticks. The real significance of the chopsticks comes forth in the accomplishment of fine and delicate gestures and in the gentle demeanour exhibited with regard to food. More important perhaps, the presence of chopsticks expose some aspects of the *Oriental world*, in particular the importance of handling things in an imaginative and non-mechanical way and of relating to others with tact and measured care. These aspects are "there" without any visual or tactile evidence.[13]

As this example makes clear, a particular world is not made up only of realities perceived by the senses. To be sure, in a given context we find various concrete things. The chopsticks are placed on the table, the table is surrounded by chairs, the chairs are moved by courteous waiters and used by well-dressed guests who take their meal and enjoy their conversation in the pleasant atmosphere of a restaurant. But the presence of a thing can also point to values, ideas, ways of living, which do not have to be present in a tangible fashion.

An old table reminds us of the presence of family members meeting at dinnertime or inviting guests or even strangers to sit down with them and share their meal. But it can also evoke the values of steadfastness, devotion, love, and the acceptance of foreigners.

A door may suggest other values. It is not merely a thing that separates the space inside and a space outside and grants the delightful experience of feeling oneself at home, in an intimate environment. Through the careful opening and closing of a door, we may become aware of the priceless experience of our freedom. Werner Herzog, in his documentary film *Little Dieter Needs to Fly*, has shown how the awareness of one's previous life experience may define the ways of seeing and handling a door. For a pilot who has been shot down in Vietnam and who has spent an amount of time in prison, the mobility of the door attests another presence; his hand opens and closes the door repeatedly and attentively and this conscious gesture of handling the door gives him evidence of a *world, in which one can move freely*. Likewise, in the eyes of Georg Simmel, the door

---

13 Barthes, *Empire of Signs*, pp. 15-18.

represents what is essential for humanity: the capacity of setting itself a *boundary*, but in such a way that it can always *step freely outside this boundary*.[14]

In a remarkable phenomenological description, Pierre Sansot brings out the features and significance of "the most modest object that one can find": the folding chair, *le pliant*.[15] Only someone truly experiencing a presence can write so colourfully and so convincingly about such a seemingly plain thing and the nature of the world that this thing brings to the fore. The folding chair is a "poetical thing;" its presence announces numerous other realities, both concrete and abstract ones. Reviewing them one by one provides the anthropologist a delightful entertainment as well as an occasion to attest the "*greatness of ordinary people*" who act with defiance against the harmful influence of the consumer society. Sitting on a folding chair, without back– and armrest, can be quite tiresome in the end. The sitting posture of the body suggests the lack of ease and the impossibility of adopting an attitude of "being comfortably settled in life." Much of a spectacle remains hidden for a person sitting on a low-slung *pliant* in midst of a dense crowd. The limited sight from the *pliant* then evokes apartments facing a poorly lit courtyard and workplaces in the back of the shop and without natural light. It is used by men and women with modest revenue who had to bend themselves (*se plier*) while working on the field. However, it is also a visible reminder of the right of ordinary people to find a place of rest at any point of the town and enjoy without restrictions the pleasures of communal life. The folding chair brings into mind the patience of the amateur angler and the communal enthusiasm of the fan sitting on the side of the road and cheering the cyclists of the Tour de France. Its size symbolizes a visible and proud dissent against the prevalent desire to produce, in our technological society, chiefly bulky objects and monumental edifices.

14 Simmel, "Bridge and Door," p. 7.
15 Sansot, *Les gens de peu*, pp. 219-23.

A little old woman, in her candour, opens her folding chair and, right there, it silences the trumpeting of our mammoths, of our large surfaces, of the immense towers, which appear, in their excess, inhuman. Its mere incongruous presence initiates a force of protest against a type of society.[16]

16 Ibid., p. 223.

I have already pointed out that, for a sensitive student, the encounter with a remarkable teacher leaves a powerful impression and most likely will determine all future interests and even life-long passions. There are singular moments when students feel a teacher's presence through her words or actions and these provoke radical changes in their outlook and attitudes.

In a book presenting the impact of influential teachers, Jacqueline de Romilly recalled a life-defining experience. The experience was brief and dazzling, lasting less than a minute. Her teacher organized the reading of *Iphigénie*, a play by Jean Racine. At one point, de Romilly had heard her teacher reciting two lines:

> *Et moi qui l'amenai, triomphante, adorée,*
> *Je m'en retournerai seule et désespérée.*
> (And I, who brought her, triumphant, adored,
> will return alone and in despair.)

The verse appeared to her in "all its force," revealing the tragic contrast between the joy and the pain of a mother departing from her daughter. The two lines and their rhymes, recited with an exquisite taste for the theatre, expressed the profound sorrow with "controlled and sovereign harmony." De Romilly sensed, in the timbre of the voice of her teacher, a miracle, a perfect fit between feeling and expression. She could never forget this presence, this "moment of grace:"

> The fact remains that, after so many years, readings and emotions, these two verses still ring in my ears, with the slightly rasping and trembling voice of our former teacher. And it was, all of the sudden,

> the discovery of all the great contrasts of the tragedy, and those of all possibilities and all the echoes of poetic expression in general. (...) The two lines had touched me, straight on, like in the paintings where revelation comes from heaven to strike a man in prayer with an infallible ray.[1]

Often the sounds produced by the voice of another leaves on us a stronger impression than the sight of an object. Sounds detach themselves from their source and pursue us. We are able to turn away from visible objects, but unable to preserve a distance between the sounds and ourselves. Colours "cling" to objects; sounds "move away" from them and "enjoy" an autonomous existence. We can easily impose our will on the visible, but not on the audible; we can close our eyes at any time, but not our ears.

Helmuth Plessner relates both the volume and impulse of rhythmically articulated sounds to the phenomenon of "insistence (*Eindringlichkeit*)."[2] No other sensory element makes its way, presses itself upon us, and soaks into us, as does a sound: it provokes resonances at levels deeper than a colour or a tactile quality. Not only strident sounds, such as a piercing cry, but also gentle melodies and poetic compositions penetrate and reach the depths of our being. Sometimes their effect is indelible. A tune or a verse, heard many years ago, can suddenly come to our consciousness and asserts its spell.

In his lessons on art, Georg Wilhelm Friedrich Hegel reflected on the privileged place that the singing voice occupies among musical activities. He clearly marked the fundamental difference between vocal and instrumental music: the voice gives an unmediated expression of our inner life, the musical instrument, in Hegel's eyes, is an indifferent mediation between feelings and the audience.

> The human voice can apprehend itself as the sounding of the soul itself, as the sound which the inner life has in its own nature for the expression of itself, an expression which it regulates directly. On the

1 De Romilly, "Diptyque," p. 50.
2 Plessner, "Anthropologie der Sinne," pp. 22-3.

> other hand, in the case of the other instruments, a vibration is set up in a body indifferent to the soul and its expression and, in virtue of its own character, more remote from these; but in song the soul rings out from its own body.[3]

Indeed, in our contacts with others, the variation, intensity and sonority of our voice immediately reveal our emotional dispositions. Hence, a certain embarrassment that we can feel when we are called upon to sing alone or even recite a poem in front of strangers. We then have the impression of being delivered, through our naked voice, to the immediate appreciation of others. From the point of view of the relationship to ourselves, singing or reciting can give rise to a similar discomfort: it can suddenly create a self-presence by putting us in the presence of our desires, our ideas and perhaps even of our most personal passions. Serge Wilfart tells us that the practice of singing can even lead to "a complete questioning of one's personality and one's vocation."[4]

What is it that becomes present in a dedicated teacher's voice? The students sense a passion, a steadfast and intense inner attitude towards a particular area of study. Passion, in Louis Lavelle's words, "is pure receptivity and consent to the inner impulsion, and total imperviousness to any voice coming from without which would seduce it out of its course."[5] Passion is expressed by a committed and generous attitude, by a willingness to sacrifice and forget oneself for the sake of a single purpose. Certainly, passion implies a concentration, the acceptance of constraints imposed by something perceived as greater than oneself. The curtailment of passion entails a certain limitation of freedom, a strain imposed by the object of passion. However, thanks to its attraction and its multiple possibilities, the object of passion procures, at the same time, the joy of being fascinated by something and the realization of one's own potentialities. In short, in passion, self-limitation and self-realization are united in a paradoxical way; it allows one to find

3 Hegel, *Aesthetics*, p. 922.
4 Wilfart, *Le chant de l'Être*, p. 129.
5 Lavelle, *The Dilemma of Narcissus*, p. 170.

fulfilment through effort, work, care, as well as to accept a certain restraint required by the object of the passion.

The contact through the auditory sense brings forth a singular experience of *auditory presence*. We might have a similar experience of presence through the *sense of touch*.

Here is a personal illustration of this experience. I was once sitting in a dentist's chair and expected the administration of an anaesthetic injection into a mandibular nerve of my face. As the needle penetrated the maxillary artery area and as I felt the acute pain, a nurse, standing beside my chair, put her right hand on my left shoulder and pressed it gently. The pressure of the hand on my shoulder lasted several seconds. It eased the discomfort and tension I felt in my whole body and took away the sharpness of the pain. More important, through her consoling touch, I felt an inward calm and peace and a tangible and unexpected presence of a hand communicating kindness and care.

It is customary to make a distinction between a cognitive and an affective touch of the hand. Doctors and nurses are usually called to touch the patient's body in a detached and impersonal manner. This touch is obviously completely different from the gentle caressing touch of a lover. The dispassionate palpitation and the amorous fondling are radically foreign to each other and offer two utterly different forms of communication and of knowledge of the body. Passing from first approach to the second by the gynaecologist prompts in the patient a profound insult and humiliation. But there is a manner of touching someone, which is neither impassive nor private; it is affectionate and sympathetic. After announcing a disagreeable news to a patient, a physician might touch carefully the arm of the patient and in this way signal his compassion and readiness to help in the future. This form of delicate touch can often produce a greater sense of soothing and serenity than can an encouraging remark. Erwin W. Straus also tells us that "A gentle touch, a slight pressure of the hand can bring me 'closer' to another person than can a glance or a word."[6] The person who is touched this

6 Straus, *The Primary World of Senses*, p. 383.

way knows that the gesture was exclusively addressed to him. While the sound of the teacher's voice reaches everyone in the class, the nurse's caring touch is intended for one person only. "The sense of touch is an exclusive sense," observes Straus correctly.[7]

There is in the nurse's touch a will to act with *modesty* in the presence of a living body. While placing the hand on someone's shoulder or elbow, the possible suspicion of invasion, of possession, and promiscuity can be avoided. The senses are held in check; they are submitted to a certain discipline, a discerning control. The chief concern of the modest hand is to welcome the other and offer, without any other sort of wish, comfort and hospitality. "Hospitality," argues Bernard Matray, "is just another word for acceptance of the other's presence."[8] While offering hospitality, the gentle touch implicitly recognizes the presence and dignity of a person in a particular situation.

The caring touch also establishes between the two persons a relation of *intimacy.* Something of the person's inner life becomes present in the act of touching. This inner life can comprise force, generosity, a feeling of sorrow and concern, an understanding of fragility and vulnerability of the human body.

The person who is carefully touched must feel not only the expression of sympathy but also the respect of his inner life. Respect is inseparable from care; it is, in fact, care in its practical, affective and vibrant manifestation. It is sustained by a fine sense of guessing correctly what is proper, decent, and appropriate, by good taste, by tact.

The act of touching evokes, for Eugène Minkowski, the virtue of tact, "an exceptional delicacy of humans" in their relations with their fellow men and women.[9] Tact is based on the willingness to preserve a distance; it respects the intimate sphere of the other person. When we say or do something tactfully, we also rely on our feelings for the right form and the convenient proportion. Tact is a sensitivity to the

7 Ibid., p. 373.
8 Matray, *La présence et le respect*, p. 57.
9 Minkowski, *Vers une cosmologie*, p. 187.

appropriate behaviour in situations where the knowledge of rules, manners, and customs does not offer sufficient guidance. What is revealed in a presence marked by tact is an attentive receptivity to the concrete person, the sense of affective nuances above all, which suits a particular need or demand. Tact translates into a particular sensitivity to the changing and unforeseen circumstances of real life, by a capacity to feel the concrete desires, worries, or aspirations of another person.

Without tact, we would hardly be able to feel the mood of people, guess their interests, their concerns, and, at the same time, to maintain a certain restraint, a suitable moderation. Because tact protects us from embarrassing intrusions and hurts, it helps us to avoid clumsy abruptness and embarrassing collision; it prevents possible uncomfortable shocks.

David Heyd pointed out that tact also requires, while maintaining a respectful attitude, the ability to identify ourselves, with our whole being, with another person and to feel his sorrow or joy:

> Only by transposing an experience to oneself can one become sensitive to the feeling of the other. And as there is no particular organ of touch, the whole body serving as sensor, so is tactful behavior the outcome of an attention to the personality and circumstances of the other, achieved by general openness and sensitivity, rather than by any specific skill or faculty, cognitive or emotional.[10]

In addition, tact plays an important role when a person spontaneously does something for a fellow human being out of kindness, when he displays solicitude. The particular awareness of the weakness and vulnerability of the other, and the receptivity for whatever the other can offer in exchange, are essential parts of true solicitude.

Beyond the voice and the touch, conveying an *auditory* and a *tactile presence*, the whole body of the person may also irradiate a presence. The characteristics of the body, the pleasant proportion of its parts play a role in the perception of the presence. Actors often

10 Heyd, "Tact: Sense, Sensitivity, and Virtue," p. 221.

have a face with harmonious or striking features and build an acting career based on their appearance. However, in the experience of presence, attractiveness is often not enough. The face must express a distinctive inner quality, a strength, a resolution, a peace, or a firm commitment to a goal. The face of a grieving person expresses a profound sadness or sorrow. That of a child finding a hidden treasure irradiates an angelic joy. How transfigured is the face of a youth reading a captivating novel. He identifies himself closely with the characters, relives their tribulations or delights, and is deeply touched by their actions and emotions. His presence springs from the thrill that he gets when a swiftly developing story unfolds with twists and turns.

A presence may be experienced when a person adopts a particular attitude towards the body. Gerd Haeffner mentions the leader's ability to impose a control over his body and, thanks to the manifestation of an inner force, to exert a positive influence on others.[11] The body suggests from the outset a certain inner disposition: for instance, the leader's firm bodily carriage hints at the strength and resolution of the person, the ability to face an opposition and to reject a humiliating compromise. An "inner attitude" is formed by values, experiences, desires, representations, and expectations and these are immediately noticeable in the manner of standing, of giving directives, of engaging in a conversation, or giving a public presentation. The self-assurance in the voice, the judicious change of the rhythm of the speech, the voluntary movements of the hands disclose an intense presence. In exerting strict control over the body, a person demonstrates the ability to hold his emotions and valuations in check and to speak and act with restraint and elegance.

A *magnetic presence* may come through even if the body is not vigorous and expansive but rather weak and adaptive. A person of old age may exude authenticity and wisdom even when his body appears feeble and fragile. It is not uncommon to visit a former teacher, who sits quietly in an armchair, withdrawn from the hustle and bustle of daily life, smiling gently, and saying very little. In these moments,

11 Haeffner, *In der Gegenwart leben*, pp. 140-1.

the silence reigning in the room communicates something that both host and visitor are unable and unwilling to put into words. The visitor is surely pleased to hear words from the host – words, which may express feelings of recognition and encouragement. However, as Louis Lavelle explained with eloquence, the silent presence offers a more complete sort of fulfilment than words.

> There is no being, whom I love, who can do without all testimony, who does not need consolation or support and for whom the words are not the attention's sensitive touch and the very gift of my presence. But it is in silence that the spiritual and total gift of myself takes place.[12]

Silence introduces to this personal offering of the gift a note of gravity and solemnity. The visitor senses that, in this moment, he is stepping out of the ordinary course of life and experiences an atmosphere of festivity and calm. Through the instauration of festive stillness, the host intimates the genuine significance of actions, relations, attitudes that fill the everyday existence. He suggests the difference between what really matters and what seems just important, what fills one with energy and what produces fatigue, what is communicated in silence and what is marked by obstreperous noisiness.

We all know that the entrance of another person into a room, in which we find ourselves alone, instantaneously modifies the way we feel: we become attentive, cheerful, energized, reserved, tense, annoyed, or irritated. We experience a very slight, hardly noticeable pressure coming out of this person. In fact, our actual state of mind and body is a reply to the modification of the atmosphere, which is produced by the entrance of the person and by the significance of his action in connection to ourselves. I have already mentioned that an atmosphere is a stirring emotional quality permeating a lived space and exerting an explicit or tacit influence on our thoughts, feelings, and actions. To detect a particular atmospheric quality means to reach beyond the factual, the objectively given: to hear, beyond the sound, and to see, beyond the visual shape. Our first impression

12 Lavelle, *La parole et l'écriture*, p. 137.

of a person occurs by virtue of such an immediate experience of a distinctive emotional quality.

We are all responsive, consciously or unconsciously, to the atmosphere created by people around us. Our behaviour is shaped, to a greater or lesser extent, by the affective tonality emerging from the persons with whom we interact, from our colleagues, our employers, or our friends. For this reason, we may enjoy playing with our friends in a music band or in a sport team or we may resent our participation in a negotiation with a unpleasant business partner. In general, a personal atmosphere permeates every sector of our life-world and influences, to a greater or lesser degree, the characteristics and outcomes of all our activities.

Occasionally, we may find ourselves in presence of an astounding atmospheric nimbus of a strong personality. We then sense a particular *aura* and, with it, a certain tonality which, like the perfume, suddenly or gradually infiltrates the whole surroundings.

Here is an example illustrating the aura of a person experienced during the rehearsal of the Berlin Philharmonic Orchestra and the response that this presence elicits. A percussionist of the orchestra tells the story in the BBC's documentary on *The Art of Conducting*:

> I had studied composition and conducting. Therefore, I always had a score open in front of me. When I had nothing to do, during my long pauses, I was just following the composition with the score. During rehearsals, one gets used to a particular sound. Suddenly I heard an entirely new sound, a fascinating sound. I asked myself what is really happening. I looked at the podium; nothing special was to be seen there. Then I looked at my colleagues and they were all staring at the entrance. In the entrance stood Furtwängler. He was just coming in. It was his personality, his appearance (*Auftreten*) alone, while he was doing nothing, which had created this incredibly beautiful sound. When person carries the sound so strongly within himself, then, he brings out the sound in others. Of course, it is the most beautiful thing that an orchestra can experience. When you notice that this person is completely open, you are invited to do the same. This is how and when you make this kind of music.[13]

13 https://www.youtube.com/watch?v=oM9ZCmyD0gM

Certain persons have this capacity to be present. They carry within themselves a particular quality, value, or knowledge. Their external appearance may be insignificant and pedestrian. Yet they irradiate something that seizes the onlooker: perhaps a life experience, a goodness, an uncommon warmth, or an artistic gift. Wilhelm Furtwängler carried music within himself. All the works he had learned and performed previously contributed to the creation of an aura around him. This aura, as we have seen in this example, created a profound effect on the orchestra's playing: a sudden transformation of way of making music. This happened by the conductor' s way of standing beside the entrance, by his fascinating presence.

Charles Munch, the famous and mesmerizing conductor of the Boston Symphony Orchestra, considered the magnetic presence an indispensable characteristic of a conductor:

> Those fifteen years of work and study do not make a conductor of a man if he is not possessed by an inner exaltation, an all-consuming flame and a magnetism that can bewitch both the musicians of his orchestra and the audience come to hear his music-making.[14]

This aura may also come forth through the movements of the *hands*. In one of his short stories entitled *Twenty-Four Hours in the Life of a Woman*, Stefan Zweig describes how the unique, beautiful and magical pair of hands make a young gambler present and how this presence holds spellbound a middle-aged woman. She is mesmerized by the eloquent hands of a young man, in which "passion broke almost tangibly from the pores of the skin." She felt drawn not only to a pair of extraordinary hands, reflecting the ebb and flow of feelings, but also to an intense and fascinating face, in which "passion showed so openly," and to the risk that an eventual encounter with this young man inevitably creates.

Seldom do we have the chance of finding ourselves in the workshop of a painter or a sculptor or in the room of a writer or a composer while they create a work of art with their hands. Musicians playing on an instrument do make possible the observation of a creating

14 Munch, *I am a Conductor*, p. 4.

hand. In a concert hall, their technical skills, feelings, liveliness, virtuosity, potential for invention, variation, and novelty, their entire experience of learning and performing become present in their hands. Some auditors watch so attentively the pianist's hand that they forget to listen to the succession of sounds. Others, however, notice, with both their ears and eyes, the presence of hands presenting a highly personal interpretation of music. The play is more personal in the sense that it leaves room for very slight deviations caused by the emotional reaction of the musician. The quality of a tone, a phrase, a rhythmic figure, the acoustic quality of the hall may, by resonance, awaken in the musician a vital impulse which, through the "intelligent presence of the ten fingers" (Jankélévitch), gives rise to a subtle change of tempo, a more marked accent or a softer or more brilliant sonority. Subtle deviations provide the performers with opportunities to make their playing livelier and emotionally charged and the listener with moments of surprise and even delight. A somewhat informed listener can distinguish between the pianist who presents the work with scrupulous fidelity and the performer who adds to his playing "an element of sympathetic participation."[15]

But such a presence appears perhaps with an even greater brilliance and persuasiveness when we see the movement of the conductor's hands. We tend to forget that the manual performance of a conductor is just as dexterous and ingenious as is that of a pianist or of a violinist. Conducting an orchestra with brio calls for just as much learning of the movements as it does for a solo instrumentalist. Virtuosity is not the privilege of a performer playing a musical instrument. Conductors are also endowed with suggestive force and virtuoso technique while leading an orchestra. The conductor's performance consists usually of drawing, with the left hand, the lines of melodies and thematic progression and of indicating, with the right hand, music's pulses, dynamics, and rhythmic structures. But the hands may also suggest to both the orchestra and the audience all the emotional elements that may have prompted the composer to create the work and, sometimes, the concrete or abstract meaning

15 Jankélévitch, *Liszt et rhapsodie*, p. 116

of the piece. Certain conductors, such as Ferenc Fricsay or Klaus Tennstedt, bring forth with their movements more than their musical expertise and conception. Their hands are the image of their artistic preferences and intuitions and of the characteristics of the communicative relations to their fellow human beings, chiefly to the members of the orchestra and to the audience. The love of music and the sense of awe may appear on the hands, as the movements take a larger amplitude and, through the music, the conductor's body seems to grow in stature. His whole being irradiates these feelings and the members of orchestra adopt them and translate them into the performance.

The aura of an artistic personality emerges in a particular fashion when a conductor plays on a piano or a harpsichord and, by lifting up one hand from the keyboard, conducts a chamber orchestra. For example, Karl Richter's playing and conducting make present both the spirit of Johann Sebastian Bach and his own personal approach to the interpretation of the musical work. For the attentive listener, Richter appears shifting from one instrument to another: the hands play on the keyboard as well as on the orchestra. And, in both cases, they produce sounds and let themselves be guided by the already produced sounds.

Indeed, music, as envisaged by the composer and set down on the score and as already played by the orchestra, defines, in part, the movements of the hand. Fascinated by the movements, the audience not only hears the music but, to a certain extent, also sees the music. Obviously, the listeners perceive a musical piece by hearing the sounds: the sounds reach them, envelop them, and, literarily, penetrate them. But intently watching the conductor's hands also contributes to their auditory experience. Their understanding and enjoyment of the piece become more complete. The politician and writer Theodor Heuss attended the philosopher Theodor Lipps's lectures in Munich. He tells us that he could read a whole thought process of the lecturer by observing his "tense facial expressions." In a similar manner, it is possible to read the conductor's conception of the work by paying close attention to his gestures and the muscle tensions of his hands. Tranquillity, daring, passion, tenderness, playfulness, admiration, and many other emotions or attitudes adopted by the conductor in relation to the music may vividly come forth in the animation of the hands.

The listeners experience a singular magical moment right after the conductor's hands has waved the final chord. "What a moment of silence," exclaims Bernard Welte, "a moment of fulfilment, of plenitude, of radiant presence."[16] It is an instant of deep contentment, as the work has reached completion and became an integrated whole. It seems as if the hands catch their breath after such a remarkable effort, as the audience remains silent for a split second before bursting into applause. All the previously heard sounds and silences, all the melodies, rhythms, and chords are concentrated in a singular and enchanting moment, in a pure and timeless present. In this moment of silence, the listeners encounter a concentration of the musical work, its presence beyond the temporal succession of tones. The musicians relish the result of tense and concentrated rehearsals and the satisfaction expressed on the face and the hands of the conductor. Both, the performance and the approval, give them an unforgettable experience of presence. "The past as the completed, the completed as the present, has thereby become free from the burden of effort, from work and worry, and from everything that goes with it."[17] There are other similar situations, in which the long and persevering effort yield a feeling of contentment and an awareness of presence. One enjoys the gratification arising from the sense of accomplishment when the last sentence of a substantial doctoral dissertation is written, the last suture of a complicated chirurgical intervention is placed, the last stich on a festive dress is sown, or the last step of an intriguing mechanical problem is solved – all these by the dexterous and creative hand.

Welte mentions the blissful moment experienced by intimate friends who, after having spent some time conversing and remaining in silence, say farewell to each other. Perhaps in no other moment they feel the presence of the other, as well as the presence of their

16 Welte, "Meditation über Zeit," p. 23.
17 Ibidem.

friendship itself, as in the moment of embracing each other and saying good-bye to each other. This moment of departing from a kindred soul offers a sense of presence of a completed and intensely lived togetherness.

This sense of presence is part of the life of those who, beyond their friendship, are in love. The poem of Vladimir Nabokov *The Swift* gives a fine illustration of their sublime experience.

> One night between sunset and river
> On the old bridge we stood, you and I.
> Will you ever forget it, I queried,
> – that particular swift that went by?
> And you answered, so earnestly: Never!
>
> And what sobs made us suddenly shiver,
> What a cry life emitted in flight!
> Till we die, till tomorrow, for ever,
> You and I on the old bridge one night.

The poem evokes three forms of presences: the old bridge over the river, mentioned twice, the passage of the bird, and the lovers' deep and lasting sentiment for each other. The bridge, as we have seen, lifts up two persons from their everyday concerns and turns them to what matters in life. As Paul D. Morris has pointed out in his fine analysis of the poem, "The spatial and temporal setting of the poem contributes to the evocation of a momentary departure from the contiguity of time."[18] Standing on a bridge and seeing the flight of a bird reminds them of the trajectory of their life, marked by an ever-lasting unity, and of the contrast between fleeting and lasting realities. Their shared deep sentiment is put in relation to the appreciation of a trifling event: the flight of a bird. Equally, the expansiveness of the love of two persons, their sense of being indissolubly and endlessly united, is juxtaposed with the contraction of time into a dense and passing moment, which will be remembered forever.

18 Morris, *Vladimir Nabokov*, p. 186.

Self-Presence

Self-presence is also an experience of presence. How can we characterize it? We may first define it as a state of mind, in which distraction and scattering of attention are absent. The distracted person is unable or unwilling to dwell on a scenery, a work of art or a useful and fine looking thing such as a well-crafted knife. The eyes or the hands do not rest with it. We may illustrate this state of inattentiveness by observing persons going through the various halls of a museum and casting a furtive glance at the paintings. Their vision jumps from one painting to another without looking at their pictorial quality, without noticing the unity of their formal elements and without being aware of the subjective thoughts, emotions or even memories that their perceptions evoke. The distracted visitors' vision does not come to rest in the act of looking at the paintings and in the attempt to understand the characteristics of their subjective reaction to the experience.

How different is the attitude of the visitor who clings to the present experience, beholds a painting with the outmost attention, and retains in memory the sensory and non-sensory content of the experience. The visitor collects himself and his vision rests awhile with the painting.

Collecting ourselves is being present to ourselves, being aware of our original, bodily presence in the world. It requires that we resist the distracting pressures coming from demands, concerns, desires, or practical plans. We are able to resist being pulled into various directions because we are aware, in this state, of our fundamental freedom: we either remain unconcerned about external demands or yield to distracting solicitations.

In simple terms, collecting ourselves means freely retiring within ourselves, going into ourselves. Eugène Minkowski calls the act of withdrawing into oneself (*se recueillir*) *interiorization*.[1] In this withdrawal, we reach "the very base," the "centre" of our being. Writing about the meaning of the verb "collecting oneself (*sich sammeln*)," Romano Guardini also emphasizes the ability of gathering up, of finding a centre in ourselves and of being united with this centre in ourselves.[2] This withdrawal into ourselves, centring in ourselves does not mean escaping from the world and cutting ourselves off from our senses, feelings, and passions and entering only to a purely spiritual or even supernatural sphere. Collecting ourselves implies being present to other human beings and to the living and unliving worlds and being aware of all the dimensions of our human life.[3] The self-presence and presence to the world in the state of gathering ourselves together remains a life-long task. To achieve this state, we have to overcome the constant temptation of distraction.

We may freely return to ourselves and unite with ourselves without any specific reason. We may sit in a quiet park, turn away from all tasks and external demands and go into the depths of our being. Or, as Karl Baier pointed out, we attain self-relation spontaneously and without warning when we find ourselves at a turning point of our life history: for example, in the moment of leaving the parental home, of commencing of a new kind of career, of learning about a severe illness, or of retiring from a professional occupation.[4] These moments have a certain duration that, thanks to their deeper importance and transforming effect, stand out with regard to the past and to the future in our personal becoming and make us stop, force us to stay in the present and withdraw into ourselves. Finding ourselves in the obligation of making a decision is another pivotal moment in our personal becoming. When we face a concrete

---

1 Minkowski, *Lived Time*, p. 106.
2 Guardini, *The Art of Praying*, pp. 13-14.
3 Baier, "Phänomenologie der Sammlung," pp. 40-2.
4 Ibid., p. 40.

instance of deciding between two excluding possibilities, we are thrown back upon ourselves and called to rely on our own resources. It is not uncommon, in this moment, of considering the prescription of an accepted and internalized convention and opposing it to the promptings of our heart. How are we going to decide when we are passionate about becoming an actor but a business organization offers us to pursue a business-oriented lucrative career path? We are forced to deal with the decision alone, however dreadful the absence of helping advice may be and whatever energy the freedom from all external influence may consume. But the startling difficulty has its reward: we are compelled to return to ourselves and to confront and discard, with courage and strength, one of the opposite possibilities alone.

Collecting oneself is eminently present in the *act of praying*. Prayer brings us face to face to ourselves as we turn away from busying ourselves with various seemingly important businesses. In his phenomenological analysis of prayer, Eugène Minkowski stressed the degree of withdrawal into oneself:

> I live the total interiorization in it. Beside this, how small a thing is reverie, which is sometimes considered as the prototype of interiorization! Prayer leads me straight off "to the very end" of this standpoint; the interiorization attains its maximum there.[5]

But, in prayer, as well as in other forms of interiorizations, we do not merely remain face to face with ourselves. We lift up ourselves above ourselves and reach out beyond this world. We "turn our eyes to the distance, toward an infinite horizon, toward a sphere beyond time and beyond space, a sphere full of grandeur and clarity but also mystery. We can then replace this mystery with the idea of an active divinity."[6]

Even if numerous persons gather in a church, each person is present to himself and prays in a state of self-collection. Although this state opens the way to a transcendent being, he is fully aware

---

5 Minkowski, *The Lived Time*, p. 106.
6 Ibid., p. 103.

of his power of leaving behind the noise, worry, and busyness of his everyday life and being in touch with the inner centre of his own being. It may happen that, in this moment of solitude, he discover within himself unsuspected insights or purposes or unanticipated devotion or adoration. Silent retreat within oneself is an essential part of prayer. As Romano Guardini pointed out, the person must turn his back to everything and "become and remain present."

> This *remaining* is difficult, for only on rare occasions do we meet with an experience so definite and compelling as to hold us and make us willing to remain for a while. Yet everything depends on this ability to stand still and to be present with full inner awareness.[7]

Music, as we have seen, creates a silence and this kind of silence may foster the disposition of praying with inner awareness. Such a silence "listens" both one's own inner life and an "infinite horizon" full of meaning. Far beyond the sounds, argues Josef Pieper, "music opens up a great, perfectly dimensioned space of silence within which, when things come about happily, a reality can dawn which ranks higher than music."[8]

In her widely read book, Anne Morrow Lindbergh, fully conscious of the constrains preventing women from paying attention to themselves, mentions prayer and other activities, which help them to resist the temptation to distraction and to foster a self-presence.

> Woman's life today is tending more and more toward the state William James describes so well in the German word, "Zerrissenheit – torn-to-pieces-hood." She cannot live perpetually in "Zerrissenheit." She will be shattered into a thousand pieces. On the contrary, she must consciously encourage those pursuits which oppose the centrifugal forces of today. Quiet time alone, contemplation, prayer, music, a centering line of thought or reading, of study or work. It can be physical or intellectual or artistic, any creative life proceeding from oneself. It need not be an enormous project or a great work. But it should be something of one's own. Arranging a bowl of flowers in the morning

---

7 Guardini, *The Art of Praying*, p. 13.
8 Pieper, "Music and Silence," p. 56.

> can give a sense of quiet in a crowded day—like writing a poem, or saying a prayer. What matters is one be for a time inwardly attentive.[9]

In all these activities, women are able to fend off distraction and achieve a self-presence.[10] Although they are primarily concerned with their activity and the concrete object (flower, poem, musical piece) of this activity, they are also "inwardly attentive" to the way they feel and think about their effort and achievement. They may qualify an activity exciting, challenging, soothing, rewarding, or straining. In all cases, they do something while being present to themselves. They purposely gather together their thoughts and emotions while engaging in a meaningful activity, thereby harmonizing their innermost perceptions with engagement with the world.

Likewise, whenever we are immersed in an activity, a conversation with a friend, we are present to ourselves. Without such an implicit or explicit presence, we would be unable to address the other persons and listen to their views, to their questions and answers. But, conversely, in order to be present to ourselves, we must first have turned away from ourselves.

This is to say that self-presence also happens when we concentrate on the accomplishment of an activity or of a perception. But this concentration succeeds to our satisfaction when we are able to relax, get rid of anxious effort, and allow our body and mind to actualize their basic potentialities. We reach both proficiency and a pleasurable awareness of ourselves as we bring together two seeming incompatibles: effort and relaxation. Then the concentration on a task as well as the performance of an action happen on their own.[11]

What we think or feel is quite often not the object of an inward observation. It may become fully conscious if an emotion or an idea suddenly gains a certain intensity: a story read in the newspaper

---

9 Morrow Lindbergh, *Gift from the Sea*, pp. 55-6.

10 Gerd Haeffner mentions similar activities: athletic performance, music making, meditation, contemplation. These activities may give rise to "an emotional experience of self-presence." "Gott der Gegenwart," p. 174.

11 Ibid., pp. 173-4. See also Huxley, "Knowledge and Understanding," pp. 64-5.

evokes painful memories, facts about our state of health create the feelings of unease or anxiety, or a novel raises questions about the trajectory of our life before or behind us. In the *Agnès Jaoui*'s film *The Taste of Others*, a wealthy and busy industrialist, brilliantly acted by Jean-Pierre Bacri, reluctantly goes to see Racine's play *Bénérice*. To his surprise, the actress playing Bénérice is his would-be English language tutor. Her beauty and moving performance draw out in him previously unknown feelings and a keen interest in the unconventional world of arts. The film depicts how this uncouth man, thanks to the fresh and astonishing contact with the foreign domain of art, deals obliquely with his newly discovered and still unpolished *inner sensitivity*. The awareness and prompting of this intense affective stirring become, within him, the dominant motivating force of his subsequent actions.

Let us now consider some of the pursuits mentioned by Anne Morrow Lindbergh. Just like in the previous chapters, concrete examples help us better elucidate the characteristics of this form of presence.

When we contemplate a thing – a tree, a door, a precious stone, or a hillside filled with grapevines – we see it in an intensive manner. We introduce a pause into our daily occupation and dwell on the observed reality without any haste and distraction and with an outmost attention. We give ourselves fully to the experience and pay attention to the details that we usually fail to notice. A precious stone may reveal shades of colour or a doorframe may expose its particular shape, proportion, and texture. We disregard the reality's functional value and we simply enjoy seeing the appearance for what it is. We find it intrinsically meaningful. Josef Pieper stressed a specific aspect of seeing things in contemplation: "it is motivated by loving acceptance, by an affectionate affirmation."[12] As we have seen, the contemplation of a landscape makes us aware, in addition to our inclination to accept and value what we see, the presence of *reverence and restraint* within ourselves. We do not want to introduce any alteration into the object of our vision. To see something, without

12 Pieper,"*Vita contemplativa* – The Contemplative Life," p. 75.

any desire of correction, is possible when we adopt a contemplative attitude.

Whether seen from a bridge or from a bank, a river, with its slow and steady movement, its smell and glimmering streams of light, captivates many beholders. The contemplation of the river may also foster in them a self-presence. The peacefully flowing water appears to be the symbol of the steady progress of their lived time: the river is "stealing onward, like the stream of life," in the words of the poet Henry Wadsworth Longfellow. It may remind them, as they come to a presence of themselves, of their own futile or fulfilling past endeavours or of their future deeds appearing with uncertain outcome.

In his poem, *The River*, Ralph Waldo Emerson notices the unchanging character of a river over time and all the things washed by the water. Nevertheless, here again, the poet's gaze is also directed inward and not merely upon the rocks, fields, flowers, and "the same blue wonder that (his) infant eye admired." What the poet sees within is the *wisdom* he gained over the years and the mature acceptance of growth, of changes and perhaps of all the deceptions and loss of illusions that he has experienced in life.

> These are the same, but I am not the same,
> But wiser than I was, and wise enough
> Not to regret the changes, tho' they cost
> Me many a sigh.

It has been emphasized that one of the chief functions of music is to counter all distracting powers assailing us each day and to establish a unity with our own being. Active music making, improvised or score-guided, creates the condition for achieving a self-presence. In a very personal passage, Gabriel Marcel mentions the inner contact happily recovered through his playing on the piano:

> How many times, at those hours when I felt cut off from myself, from my profound roots, and as it were absorbed by everyday affairs, then as soon as I sat down at the piano and let my hands wonder over the keys, have I not had the physical feeling of sails swelled in the wind, of a heart starting to beat once again. (…) As if music was able to restore

> myself to myself in an independent world, in which, a few seconds earlier, my soul dragged itself along, without any communication with a sort of spiritual no man's land, a sordid and desolate suburb.[13]

The 'spiritual no man's land' may take various forms. One of its manifestations is the accomplishment of mental or physical action without any emotional investment. There is a clear separation of thought and feeling, reality and fantasy, rational analysis and emotional expression. The approach to persons and things becomes increasingly cerebral and abstract. The affective resonance to a performance and its concomitant enthralling awareness are absent. The unhurried improvisation on a musical instrument is, then, one of the ways of finding the necessary leisure to connect an activity with feelings and, at the same time, to become "inwardly attentive."

Music gives us occasions to recapture what has been ignored or willingly excluded from our daily existence. Undeniably, for many amateur music lovers, participation in a choir or the spontaneous singing during a festive gathering offers them the desired occasion to recover the contact with their *feelings*, unlock them, give them an expressive form and thus promote an inner attitude that resists the prevalent habit of ignoring or forgetting what is truly one's own and what enhances a sense of uniqueness and authentic self.

Paul Ricoeur contends that each musical work puts us in touch with a particular feeling that we are unable to define with a single word and that we welcome as an enrichment of our emotional life:

> Music creates for us feelings that have no names; it expands our emotional space, it opens up in us a region where absolutely original feelings can occur. When we listen to *such* music, we enter into a region of the soul which cannot be explored otherwise than by listening to *this* particular piece. Each work is authentically a modality of the soul, a modulation of the soul.[14]

---

13 Marcel, "La musique et le règne de l'esprit," p. 58.
14 Ricoeur, *La critique et la conviction,* p. 261.

The pursuit of active singing or playing on an instrument or attentive listening to a live performance provides the best conditions for experiencing the intensity and novelty of a particular feeling and its immediate enjoyment. Singing has the power to unlock and illuminate the obscure recesses and unexplored depths of our being. While singing, we may become aware of the ideas, desires, and even conflicting passions that we have carried in ourselves and whose existence has been concealed from us. Anthony Storr emphasizes a singular benefit of singing and active music making:

> I have a lurking suspicion that music may be especially important to people who are somewhat alienated from the body, because playing an instrument, singing, or simply listening to music puts them in touch with their physical being in ways unmatched by reading poetry or by looking at beautiful objects.[15]

I have just mentioned the gift of singing while taking part in a festive gathering. Here the rituals, ceremonies, songs, and dances are performed, conversations conducted, meals and drinks consumed in an atmosphere of calmness and cheerfulness. A feast celebrated in an intimate setting also invites the participants to return to themselves and to see the everyday under new light. All these activities allow them to come to a self-presence: to contemplate themselves and their everyday occupations as it were from a distance and in different perspective and realize the weight and value of their activities and relations, in both professional and private settings. Karl Kerényi observed "The feast reveals the *meaning of everyday existence*, the essence of the things that surround people and the forces that work in their lives."[16]

What are these forces that, in a festive gathering, may come to our awareness? I would suggest that, in these moments, we are able to get in touch with the force of our particular *vocation*. We sense in ourselves a previously adopted purpose which seamlessly fits to our own particular situation and to the gifts we already possess. Our

---

15 Storr, *Music and the Mind*, p. 149.

16 Kerényi, "Vom Wesen des Festes," p. 74. The italics are from me.

vocation fills us with a consciousness of dignity, greatness, courage, and responsibility. In part, a vocation makes us a person, unique and distinct from everyone else. We feel it as an inner impulsion or call to work efficiently and creatively in a particular domain – education, art, science, agriculture, business, or any other useful and gratifying endeavour. We discover it thanks to restless questioning and attentive listening to our deepest yearnings. Then, without prior notice, a powerful sense of disquiet becomes palpable within us and indicates a more or less discernible avenue. Perhaps the presence of this direction asserts itself even more forcefully when we decide to ignore it or cast it aside. Paul-Louis Landsberg was doubtless right when he asserted, "we are able to run away from our calling, but we are never able to suppress the power of its elusive presence."[17]

The awareness of a vocation as an inner force may remind us of encounters with a few remarkable persons who, due to their perceived qualities, values, and achievements, exert a profound and transformative influence on us. We recognize evidence of physical or intellectual excellence – knowledge, wisdom, strength, resilience – in a person on the basis of an intuitive knowledge of competence. Their benign and inspiring influence consists of clarifying our abilities and skills while also indicating ways for us to use them effectively. They are models who suggest a vocation and articulate the nature of this vocation. They show us what we can undertake and how we can find fulfilment in the development and exercise of our talents. The force of their callings are fundamentally placid, seldom adamant. Models leave us to pursue or reject our specific vocation and to exercise or leave dormant our particular gifts and powers.[18]

We identify ourselves truly with a vocation when we express it in concrete performances. Until we give lessons in a classroom or make tables or chairs in a workshop, we have merely a vague idea about our calling of being a teacher or a carpenter. First, we anxiously ask questions about its validity. Later, as, day after day, we complete our work and notice with satisfaction suitable outcomes, we come to

17 Landsberg, *Problèmes du personnalisme*, p. 226.
18 Scheler, "The Forms of Knowledge and Culture," pp. 32–4.

realize with clarity, courage, and conviction what we are and what we are called to do. This realization may happen when we expect it the least. We are present to ourselves through the completion of the task. Thomas Carlyle gave a pithy formulation of this self-presence achieved through deeds in these terms:

> A certain inarticulate Self-consciousness dwells dimly in us; which only our Works can render articulate and decisively discernible. Our Works are the mirror wherein the spirit first sees its natural lineaments. Hence, too, the folly of the impossible Precept, *Know thyself*; till it be translated into this partially possible one, *Know what thou canst work at.*[19]

To be sure, as I have pointed out, effort and relaxation are needed to work with proficiency. When we achieve a subtle and efficient combination of the two, we execute the appropriate movements without conscious control and effort and feel with pleasure our flawlessly functioning body. In our daily life, in a state of well-being, we tend to identify ourselves with whatever lies beyond and ahead of us. Whether concerned with our projects, tasks, or goals, we may recognize that our body supports our intentions, yet we perceive ourselves as if we were "beyond it." The body is lived as a silent, dynamic, and reliable support of our undertaking. If, on the other hand, we become tired, exhausted or ill, our body appears in its materiality as an obstacle, making impossible the fulfilment of our plans. But when we have a pleasant perception of our activity and the accomplishment of the movements becomes an intrinsically rewarding experience, we are in presence of our body; we sense it with delight from inside, as it were, as a source of all sorts of possible autonomous and intelligent performances. The body is intelligent because it has the capacity for adaptation, of anticipation, of imitation, and of invention. For example, reaching for an object, the hand projects tactile sensations that correspond not only to the actual sensations, but also to the actual approaching movement. The bodily capabilities are sensed from within; they allow the performers

19 Carlyle, *Sartor Resartus*, p. 126.

(workers, artists, athletes) to move with ease and confidence. In addition, as I have already pointed out, the body can take on various forms in accordance with a situation in which a task presents itself. It can take a working figure, a resting figure, a loving figure, an athletic figure, etc. Michel Serres is right to define the body by its capabilities and by the possible figures and movements waiting to be actualized. "The body is not real, but virtual, potential. It can. (*Il peut.*) It is almost its definition."[20] The body's potentials for adjustment, correction, and invention are lived and felt as a presence of available and creative capabilities, producing the most diverse figures.

The actual skilfully executed movements may provide a pleasurable self-presence. When we are no longer concerned with the particular purpose and result of our activity, we may take pleasure in the accomplishment of complex movements and in the awareness of doing something well. For example, when we enthusiastically cut, chisel, trim, or split wood. Konrad Lorenz claims that "the joy of functioning is a blessing for the working human."[21] Without such an enjoyable self-presence, our everyday life and work would be much duller and colourless.

We are also present to ourselves when we accomplish a good action without voluntary decision, without lengthy and purposeful planning. We might, spontaneously and without deliberation, provide help to a fellow human being in state of need. *Spontaneous actions* are deeply personal. They make us aware, without any distortion, of our intimate feelings and intuitions. Because they are not prompted by either command or constraint, they can be considered as the genuine manifestation of our "interior life." The liberation from external influences may happen while we are involved in a habitual action. First, we might carefully follow conventions, representations, or expectations, and, suddenly, without warning, we sense in ourselves a readiness or an impulse to go beyond what we have previously accomplished or envisaged.

---

20 Serres, *Corps*, p. 17.
21 Lorenz, *The Waning of Humaness*, p. 143.

But such a spontaneous self-presence may be achieved when we exert a control over and against our own impulses, emotions, passions, or desires. It may happen that we decide not to pay back injury with injury, not to correct someone making a mistake or advancing a false statement, or not to give our assent to a commonly accepted but destructive ideology. We impose on ourselves a spontaneous self-command by not acting or by doing the opposite of what we supposed or intended to do. We become conscious of the presence of both an inner impulse to do something (take revenge or attack an opponent) and an ability to impose a restraint over this affective impulse, resulting in either refraining from action or transforming the nature and meaning of the action. In his *Ethics*, Nicolaï Hartmann underscored the positive aspect of the moral strength of *self-control*: it does not consists of destroying impulses, but of holding them in check and guiding them into the right direction. Deploying power over impulses is cultivating affective life itself, "a kind of setting of one's house in order."[22] Self-control should aim at the refinement, enrichment, education, and enhancement of one's life in all its dimensions.

Once again, we may evoke here the virtue of *litote,* the very quality that we have already praised in relation to the preservation of the ecological balance of a landscape. *Litote* is the conscious will to hold back something within in the exhibition of the virtues of restraint, modesty, and discretion. The exercise of appropriate restraint in the presence of an excessive and vehement inner impulse is the guarantor of refinement and elegance in our daily life. We display the virtue of *litote* when we create something without deliberately displaying all that we know and are able to do. Michel Serres goes as far as to state: "Possessing a power and not using it, this is the beginning of wisdom. And of civilization."[23]

In this retreat within ourselves, we may find our own force or vitality of undertaking and doing something worthwhile in our lives. This force within us is never without an object, which asserts itself

22 Hartmann, *Ethics*, vol. II, p. 251.
23 Serres, *Le Tiers-Instruit*, p. 192.

with its own value and meaning. The force that motivates each person to attain an end cannot be fully apprehended in an isolated form. It is only through its object that it reveals itself. The object indicates what gives weight to our existence, what moves us in our everyday lives, what fuels the inner force in us. In the stillness of prayer, while we are in contact with our own depth, this inner force and its object reveal themselves in their genuine forms. The object can be a family member, a friend, a public figure, an ideal state such as peace or prosperity, a virtue such as loyalty or disinterestedness or even, or an attitude such as curiosity, passion, or acceptance of uncertainties with regard to a new enterprise. The inner force, nurtured by these objects, defines the course of our lives and the decisions taken in the face of emerging possibilities. Its presence during the tranquil moments of contact with oneself may lead to its deepening and strengthening and its eventual manifestations in words and actions.

In his autobiographical book *Surprised by Joy*, C.S. Lewis evokes the beneficial company of his friend, Arthur Greeves, who taught him to see intensely the beauty of ordinary things: a vegetable to be cooked, a farmyard "in its mid-morning solitude," the movement of a grey cat "squeezing its way under the barn door," or the " bent old woman with a wrinkled, motherly face coming back with an empty bucket from the pigstye." He also learned to be attentive to the various elements of a peaceful landscape during the time of sunset.[1]

Similarly to C.S. Lewis's learning experience, our attention may become more focused and more fixed by a friend who invites us to look carefully at an object. But in order to maintain our attentive and inquisitive interest, the object has to, in some way, speak to us. The object will eminently speak to us when it reveals its presence. The experience of presence achieves what C.S. Lewis's friend succeeded in his teaching: it teaches us to pay attention to all the common things that usually fall outside our visual awareness.

It often happens, as Ludwig Wittgenstein observed, "the aspects of things that are most important for us are hidden because of their simplicity and familiarity."[2] Familiarity makes our vision selective, concentrating on some specific areas of our living environment and thoughtlessly overlooking many others. Things respond flawlessly to our momentary practical needs and do not require any particular attention on our part.

But when things assert their presence, they do not necessarily undergo some transformation and become suddenly strange and

1 Lewis, *Surprised by Joy*, pp. 193-4.
2 Wittgenstein, *Philosophical Investigations*, p. 50.

unfamiliar realities. Rather they keep their familiar character and, as such, they may impose themselves with a particular intensity and force. A new atmosphere created by the loss of electricity or the fading of daylight, for instance, may force upon us a cessation of our activity. Suddenly, things appear indifferent to the usage we make of them. The window does not seem to call us to approach it and to enjoy a view outside the room. An armchair does not invite us to sit down and read a book in comfort. Yet their indifference triggers our interest and elicits our attention. They seem to exert a fine pressure on us; we feel as if a foreign body were putting a lightweight on our own body. It is through this fine bodily tension that we notice that the thing, previously ignored, is now *there*, quite independently of our practical intention. It is in this way that a lamp, a bookshelf or a mirror on the wall asserts its insistent presence.[3]

We may feel a gentle pressure on our body even when we do not see a person or a thing. We feel as if someone or something was there near to us, exerting a touch on our shoulder or arm or head. Timothy Beardsworth holds that "The essence of the phenomenon is affective: a person may feel comforted, guided, reassured by a gentle pressure… the "touch" expresses an affective relation."[4] The gentle pressure elicits an immediate and powerful emotional response of inward peace, happiness, courage, or comfort.

It may happen, however, that we are unable or unwilling to bear the pressing presence of an object. We cease to behold the beauty of a landscape or the warmth and radiance of a smiling friend or the view of a sublime Madonna by Botticelli or Fra Lippi or even the simple and obvious objects of the workaday world. We look away and seek the contact with a distracting device or simply hastily move away from the magnificent painting and throw cursory looks on other portraits. Nowadays many tourists are eager to take a photograph of a magnificent landscape. Weeks after their journey, they are pleased to talk about the scenery that they, in truth, failed to see. Restlessness or impatience may cause their unwillingness

3 Leimeiszter, "Jelen lenni (To Be Present)," p. 65.
4 Beardsworth, *A Sense of Presence*, p. 108.

to abide by the pressure of a presence. Some presences may evoke painful memories. Others provide a fascination for the imagination but not an enjoyment of the direct perception of a concrete reality. There are various technological means which prevent someone from being present and noticing the presence of objects and people. There are individuals who have relied so often on the images presented by their electronic devices that they have gradually become insensitive to the presence of the "real thing." They have unlearned to pay attention to the pressing presence of a sublime scenery and seek by various means to flee the impact of their sensory encounter. They prefer to see images, which can be reproduced at will. They do not want to see what happens around them in the present and what can be pleasurably contemplated for a short duration of time.

In his book *The Image*, the American historian Daniel J. Boorstin wrote a still relevant chapter on the lost art of travel. The 20th century brought a significant change in the human desire to visit foreign destinations: "it was the decline of the traveller and the rise of the tourist."[5] "Thus foreign travel ceased to be an activity – an experience, an undertaking – and instead became a commodity."[6] I would add to this observation that foreign travels ceased to be an occasion to dwell in the present and to look, undisturbed and without haste, at a tree, a fountain or even a local vendor sitting peacefully in front of his shop. Tourists have seldom the chance to linger in some unfamiliar place. They tend to shy away from what is diverse and foreign and are satisfied with the familiar lodging, meals, and routes to follow. They expect to experience the homogenized and reassuring continuity of their usual living conditions in a foreign environment. More important, the tours they purchase have an insulating effect: unless they purchase souvenirs from a store, they hardly ever create contacts with natives and their surroundings. Guidebooks tell tourists how to find "iconic sights," "cutting-edge architecture," "chic restaurants," and how to avoid the need to consult local citizens and how not to venture to less chartered areas

5 Boorstin, "From Traveller to Tourist," p. 93.
6 Ibid., p. 94

of a city or of a land. The most formidable and effective insulating device today is, of course, the mobile phone. Its usage impedes them from marvelling, for long minutes, at houses, parks, streets, or landscapes. These realities will be viewed, back home in a cosy environment, as a collection of abstract and rather empty snapshots. The tourist walking around with a mobile phone in their hand will look at the scenery without the sensory experience of having been there. Because they have not met life around them immediately, the memory of their experience lacks vividness and emotional intensity. The purpose of travel is to provide thrilling contacts with concrete things, which may bring an experience of presence. The purpose of tourism is to distract and to deviate the person from the path leading to a presence.

Still, if we are willing to remain genuinely open to surrounding realities, presences have the power to prompt intense and fixed attention. They make us pause and dwell on things. They invite us stay in the present, with our full sensory alertness, and put aside our urgent plans and our impelling desires and preoccupations with the future. They tend to restrain our propensity for impatience, digression, and distraction. The feelings of unease and restlessness give way to a sense of inner peace and attentiveness.

While attending something, we introduce a pause in our everyday life and establish a more sustained contact with a part of the surrounding realities. We stop at something, at a painting, at a tree, or at a waterfall. Stopping means putting aside the realization of projects and closing the door to all other solicitations. When we stop at a tree, we prefer it, for the moment, to the rest of the objects and we choose to make its particular aspects present. While we look at the tree, we isolate it from all the other elements of the scenery. We may ignore the buzzing bees, the green grass, the blue sky and the rest.

The first and last stanzas of Wordsworth's poem *The Solitary Reaper* emphasizes the obligation of stopping and abiding by the presence of a reaping and singing girl.

Behold her, single in the field,
Yon solitary Highland Lass!
Reaping and singing by herself;
Stop here, or gently pass!
Alone she cuts and binds the grain,
And sings a melancholy strain;
O listen! for the Vale profound
Is overflowing with the sound.

Whate'er the theme, the Maiden sang
As if her song could have no ending;
I saw her singing at her work,
And o'er the sickle bending;—
I listened, motionless and still;
And, as I mounted up the hill,
The music in my heart I bore,
Long after it was heard no more.

To appreciate the maiden's song, the poet perceived the sound impressions as they presented themselves to his consciousness. But this experience of listening to the song still reminds us that motionless attention remains open to distractions. His listening is pulled away from the present impression. In truth, this relation between attention and distraction is not only unavoidable, but is also, to a certain extent, necessary. If we focus strenuously on one object, a tone, a colour, a word, even an idea, this object tends to fade away. Our attention cannot be rigidly devoted to a single object. In order to remain sensitive to the sadness of the song, the attentive ear has to remain susceptible to brief and periodic distractions. "Attention," writes Minkowski, "to remain alive, must necessarily include an activity analogous to it, but non-attentive."[7] In the absence of repeated and short spans of distraction, helping us to keep our perception fresh and dynamic, our attention would lose its vitality and even its sense of reality.

Attentive listening to a musical piece has always been considered a demanding task. The presence of music is fragmentary because we hear only part of the work: tones appear and disappear. If the tones did not fade away and were not continually replaced, we would be

7 Minkowski, *Vers une cosmologie*, p. 102.

exposed not to music, but to a constant noise. Tones, once sounded, must give way to other tones. The rhythm and dynamic shifts define their extent in duration and the manner in which they leave the foreground. They can swell or diminish, sound soft or loud, just as they can announce themselves delicately or depart abruptly. To hear music, we must be able to reach beyond the tone that we hear: we must be distracted from the actual tone and open ourselves to the past and the future and to bring together into a coherent whole the tones that arrive, last, and then depart. As the music progresses, we establish relationships between the current audible moment and other moments, preceding or subsequent ones, marked by the absence of the audible. To hear music, we need both attention and distraction. We relate the absent tones, by which we are distracted, to the present tones, to which we remain attentive. In fact, our listening to music is the result of bringing together attention and distraction.

Even if we introduce a halt in our activities and make an effort to perceive things correctly, omissions and projections, prompted by desires, habits, early experiences, often cause inaccurate perception. Anxiety or nervousness tend to cause superficial, fleeting, approximate perception. Gerd Haeffner is right to state, "Opening oneself to the presence of reality is a task that has to contend with numerous emotional obstacles."[8]

The source of the difficulty may lie not in our inability to open ourselves to possible presences. There are instances when the reality we perceive is just unbearable. We do not have the strength to withstand the overpowering energy of an action or of an event. An unforeseen and exceptional gesture of recognition may exert such a strong effect on us that we tend to turn away from the encounter by making a joke of it or by redirecting the recognition towards other persons.

Presences also teach us to see things in relation to other things. Just like C.S. Lewis, we see the presence of an enclave or field together with rock outcroppings and bushes, the small pond with the chimney and windows of a cottage. We become attentive to the

8 Haeffner, *In der Gegenwart leben*, p. 147.

rhythm of patterns and designs, the interplay of lights and shadows. As we see a great variety of shapes, colours, figures, lines, contours, and figures, we try to understand how the perceived forms contrast and complement each other.

A presence not only pulls our attention towards something but also directs it towards its multiple details. The perception of details is made possible by the magnetic presence itself and not by the conscious choice of our will. Because our attentive perception becomes more alert and more sensitive, we find ourselves guided from one form to another and one detail to another. We identify the nature of a detail and recognize its connection to another detail. Our contemplation becomes "meditative," in the words of Gerd Haeffner, when it is enhanced and deepened by an attitude of calm patience and genuine childlike interest.[9]

This placid and inquisitive look at a human action is manifest in Victor Hugo's poem *Saison des semailles. Le soir*. The. The poet apprehends with a meditative attention (*je médite, obscure témoin*) the dense and vivid presence of the farmer: his slow marching advance, his gradually lengthening shadow, the repeated sowing motion of his hand, and the twilight that sets in over him. All of these elements of customary farm work touch the poet's intimate being and create a deeply moving experience of presence.

C'est le moment crépusculaire.
J'admire, assis sous un portail,
Ce reste de jour dont s'éclaire
La dernière heure du travail.

Dans les terres, de nuits baignées,
Je contemple, ému, les haillons
D'un vieillard qui jette à poignées
La moisson future aux sillons.

Sa haute silhouette noire
Domine les profonds labours.
On sent à quel point il doit croire
A la fuite utile des jours.

---

9 Haeffner, "Präsenz," p. 131.

Il marche dans la plaine immense,
Va, vient, lance la graine au loin,
Rouvre sa main, et recommence,
Et je médite, obscur témoin,

Pendant que, déployant ses voiles,
L'ombre, où se mêle une rumeur,
Semble élargir jusqu'aux étoiles
Le geste auguste du semeur.

In Toru Dutt's translation:

Sitting in a porchway cool,
Sunlight, I see, dying fast,
Twilight hastens on to rule.
Working hours have well-nigh past.

Shadows run across the lands:
But a sower lingers still,
Old, in rags, he patient stands.
Looking on, I feel a thrill.

Black and high, his silhouette
Dominates the furrows deep!
Now to sow the task is set.
Soon shall come a time to reap.

Marches he along the plain
To and fro, and scatters wide
From his hands the precious grain;
Muse I, as I see him stride.

Darkness deepens. Fades the light.
Now his gestures to mine eyes
Are august; and strange, -- his height
Seems to touch the starry skies.

I have already mentioned that presences put us in touch with the affective tonality of a reality. Just like children who play with pieces of wood, stones, clothes, the attentive look can detect in a thing its emotional tone: the sadness of a tree, the loneliness of a field, or the inhabitability of a public place. In our experience marked by attention, we bring out more fully the affective quality from the perceived reality, lend expressiveness and animation to the inexpressive and the inanimate, and, to a certain degree, "personalize" the non-personal.[10]

It may be that, as we look at the bright wheat field or listen to the bell's ringing tones, we dwell on what we see or hear. There are objects, an old oak tree or a wise teacher, which impose themselves immediately as a presence. There are other simple and inconspicuous realities, which arrest our attention, but trigger their appeal slowly, only after a careful perusal of their various aspects. Our patient observation yields then to an experience of presence. We may then be invited to *savour* their visible appearance. Just like the act of attending something, savouring involves a pause, even a duration, since we must, on a sensory level, take our time to appreciate the flavour of the object. Savouring something is not merely a sensory contact with foods or drinks. It is also a particular human attitude, as Eugène Minkowski has shown in his phenomenological studies on the various sensory contact with things and persons.[11] When we savour the charm of a village or the peace of the countryside, we allow the impressions to act on us, without any haste and any attempt at analysis or criticism. We can thus savour a poem or a concerto, abandoning ourselves unreservedly to the succession of words or sounds. This attitude has a temporary character since we forget the past and the future and remain satisfied, for as long as possible, with a slice of the present. What distinguishes attention from savouring is the more intimate nature of the relationship. There is no impartial objectivity in the act of savouring something. When we linger over something to savour the "flavour" of a presence, we adopt a particular attitude towards the world by establishing a broader, more widespread, more alive, more intimate contact with it and by reducing the distance which separates us from the favoured object.

The act of savouring something brings into the fore another important element of experiencing a presence. When we savour a summer evening beside a lake, we are in presence of something

10 Beardsworth, *A Sense of Presence*, p. 85.
11 Minkowski, *Vers une cosmologie*, pp. 206-16

having a meaning in itself, regardless of the possible functional value of seeing a sunset and breathing the fresh air. We genuinely enjoy a walk in a lovely park for its own sake, independently of the possible or real benefit we might draw from such activity. The peaceful scenery that we contemplate asserts itself with its own being and is detached from any pragmatic or even aesthetic interest. It is treated as a genuine present.

Can this present be granted to us if we consider a thing with our will? Gerd Haeffner tells us that the will may occasionally play a role in our experience of presence. We may deliberately mould our attention by adopting a respectful attitude towards things and persons. "The soul of such interaction is an attitude: the willingness to let the other be in their being and in their presence. This letting be is an acceptance; one renounces the desire to get rid of someone or something or to prevent it from arising."[12] Presences invite us to recognize the inherent dignity of people and things, their independence from any possible desire to alter them or to put them into an inappropriate context. It also means that they have something to say to us and that we need to approach them with care and respect and to pay attention to their language, to what they express with their own being.

When we receive a present, we feel enriched and enchanted by the experience. How can we characterize the subjective experience of enchantment? The various kinds of experience of presence have something in common: they allow us to live fully in the present and enjoy, without haste and concern, whatever the object of our perception offers. We are able to step out of our habitual everyday world, characterized by relations of means and ends, in which we undertake actions for the sake of some useful results. Thanks to our ability of stopping at a landscape or a work of art and of becoming absorbed in it, the present is no longer viewed as a stepping-stone to some distant goals. The continuous flow of lived time lapses and the objects of our activity are valued for their own sake.

---

12 Haeffner, "Gott der Gegenwart," p. 172.

For instance, once again, the attentive listening to a musical performance, notwithstanding the evanescent nature of the sounds and the inevitable distraction, helps us to experience a certain extension and density of our present. Wilhelm Weischedel sees in the temporal and temporalizing order of music a prominent source of "profound happiness."[13] It is aroused by the duration of an order and the awareness of the rhythmic succession of tones, which does not abolish the experience of pure present. As the ephemeral acquires duration and the absent becomes present, we are offered an opportunity to gain a foothold in time. I would add that the auditory perception of such an order also provides the beneficial liberation from instrumental and utilitarian concerns. For music, the coming and going of tones, is "pure play" (Bernard Welte), allowing us to free ourselves from the burden of our prosaic daily life and to enjoy the sublime sensation of relaxation, of interruption from work, and of momentary forgetfulness of the useful. Thanks to our ability to dwell on something, music gives us the joy of the present, the joy of coming into contact with an ordered work of art that knows no useful purpose.

The reader recalls the experience of Antal Szerb sitting alone on the third peak of the Monte Titiano. The writer spoke of replete happiness, missing nothing and being entirely content with being alone and serene. Before this felicitous moment, while finding himself in the company of tourists, in hotels, restaurants, on trains, he experienced a pervading sense of uneasiness and restlessness. But, at the foot of the Third Tower, he enjoyed his solitary happiness, the awareness of a state of "self-completeness." He wanted to preserve this delightful condition from any kind of influence that may be exercised by a state or a set of beliefs.

A somehow similar experience was lived by Annie Dillard as she was pulling into at a gas station in the state of Virginia. She was completely alone and decided not to drive away immediately from the rather prosaic spot. She stopped, looked around and contemplated the sublime scenery. Her view of the landscape, composed of "a

13 Weischedel, "Fragment über die Musik," p. 206.

low hill, trembling in yellow brome," and "an enormous mountain ridge, forested, alive and awesome with brilliant blown lights," granted her an extraordinary experience of presence. "I have never seen anything so tremulous and alive." Dillard described how the mountain animated before her appreciating eyes, how the lights, shadows, colours, and forms changed, contracted, expanded, and sprouted, how the "mountain loom(ed) miles closer," and how this rich appearance transformed her. She exclaimed in delight: "I am more alive than all the world."[14]

The sensuous enjoyment of feeling fully alive is perhaps the most important gift of experiencing presences. What does it mean being fully alive? I would say being fully alive is being able to wonder at the things around us, to apprehend them through strong and free emotions, and to have fun while treating them playfully.

When we enjoy being fully alive, we turn to the world with receptiveness and "energetic readiness" to perceive objects around us with attention and wonder. The source of wonder is the awareness of the *being* of a thing. We wonder about the fact the things *are*, assert their presence and offer themselves to our senses.[15]

We wonder when we carefully and questioningly peruse a landscape and realize that it ceases to appear in its banal and indifferent obviousness; it suddenly and directly comes into our view with its animated presence. We wonder when we realize that a seemingly trivial thing suddenly presents itself with an unusual intensity. We gradually go beyond the general traits to reach that which constitutes the uniqueness of the thing.

In the moment of wondering, we stand still, pay close attention to something or somebody that is no longer obvious, seek to reach beyond what is apparent and taken for granted. We also reach beyond what is opaque and obscure, and puzzle over the possibility of hidden and still unknown facts or causes.

We do this because we become genuinely interested in the peculiar characteristics of an object; we are captivated by its attractive

14 Dillard, *Pilgrim at Tinker Creek*, p. 79.
15 Haeffner, "Staunen," pp. 24-5.

presence. By asking a question, we introduce a break in the regular flow of our physical or mental activities and achieve the desired distance from ourselves and from the persons or things we would like to know better. Wonder implies being-touched by an event, being-stirred deeply by the novelty and strangeness of what, in this event, appears to another onlooker common and self-evident. Hence, wonder is an experience of enlargement; it stimulates our intellect, imagination, and feeling.

Being alive also means that we are fully in touch with our emotions. When Szerb gives attention to his surge of happiness, he senses his own uniqueness, his authentic, independent, and true self. G.K. Chesterton is right to say that the emotion, "the irreductible minimum, the indestructible germ," is "the only existing thing which is true."[16] When we recognize the active force of an emotion in relation to the objects of our surroundings, we become mindful that our identity is not merely a process of self-distancing and self-controlling thought. The more we are in touch with our emotions, the more we become aware of our own identity and the more we are able to shape our individual lives. There is an essential connection between emotional self-awareness and autonomy: we are able to manifest our relative autonomy in the world when we become attentive to and act upon the full range of our emotions.

Equally, there is a connection between an emotional state and the appearance of our surroundings. The presence of an object, as we have seen, triggers a singular feeling of peace and happiness. This free and enchanting feeling, on its turn, lends to an object glow, intensity, and purity. In Beardsworth's book, one contributor speaks of "the incredible definition of everything."[17] When things appear with a greater clarity and vivacity, the attentive receptiveness also becomes sharper. Another contributor recognizes this powerful effect and highlights "a consequent greater openness towards all and a widening concern for others."[18]

---

16 Chesterton, *Twelve types*, p. 7.
17 Beardsworth, *A Sense of Presence*, p. 61.
18 Ibid., pp. 60-1.

In his essay on Alexander Pushkin, Vladimir Nabokov made the following comment on the source of art:

> And how entertaining life becomes when you put yourself into a frame of mind where the most elementary things reveal their unique luster. You walk, you pause, you watch people pass, and then the hunt begins. And when you notice a child in the street transfixed by the sight of some incident that he will surely recall one day, you have the sense of being time's accomplice, for you see that child storing away a future recollection that already seems to adorn him. And how enormous is the world![19]

How can the most elementary thing, such as a chair or a bed, appearing with a glowing presence, become entertaining? How can it provide amusement, fun, and also wonder? What kind of the frame of mind elicits this change from a thing? I would say that the adoption of a light-hearted attitude of play make things appear as if they were glowing a pure radiance. The playful frame of mind gives us a ticket to enter another world – a world presenting expressive and dynamic qualities – which we usually fail to see or neglect to see. We envisage things with a new frame of mind and creatively free ourselves from all old assumptions about them. We gain this freedom not merely through subjective decision. We catch sight of a new world partly because the objects of our everyday life – tables, utensils, pencils, lamps – reveal their animated and enticing presence and invite us to approach them differently. This new world is entertaining and fun because things acquire a new significance; they intimate new possibilities of approach and arrangement; their perceived dynamic qualities may undergo further innovative transformation as our playful approach fosters a reciprocal and creative interaction between the animated thing and us. Our sense of enjoyment and light-heartedness nurtures our desire to treat things around us playfully and sustains the eventual repetition of our inventive and altering activity.

19 Nabokov, "Pushkin, or the True and Seemingly True (Essay 1937)," p. 132.

Bibliography

Baier, Karl, "Phänomenologie der Sammlung," *Daseinanalyse 15* (1998), pp. 36-44.

Barthes, Roland, *Empire of Signs*, trans. by Richard Howard (New York: Hill and Wang, The Noonday Press, 1992).

Beardsworth, Timothy, *A Sense of Presence: The phenomenology of certain kinds of visionary and ecstatic experience, based on thousand contemporary first-hand accounts* (Oxford: The Religious Experience Research Unit, 1977).

Berg, Jan Hendrick van den, *The Changing Nature of Man: Introduction to a Historical Psychology (Metabletica)*, trans. by H. F. Croes (New York: W. W. Norton, 1983).

– *A Different Existence. Principles of Phenomenological Psychopathology* (Pittsburgh: Duquesne University Press, 1995).

– "The Meaning of Being-Ill," *Medische Antropologie* 24 (2012), pp. 383-91.

Boorstin, Daniel Joseph, "From Traveller to Tourist: The Lost Art of Travel," in *The Image or What Happened to the American Dream* (Harmondsworth: Penguin Books, 1963), pp. 86-125.

Buytendijk, Frederik Jacobus Johannes, "Das menschliche Spielen," in Hans-Georg Gadamer, Paul Vogler, eds., *Kulturanthropologie* (Stuttgart: Georg Thieme Verlag, 1973), pp. 88–122.

Carlyle, Thomas, *Sartor Resartus* (Oxford: Oxford University Press, 2008).

Chesterton, Gilbert Keith, "Lamp-Posts," in *The Uses of Diversity: A Book of Essay*, 3rd edn. (London: Methuen & Co. 1921), pp. 7-12.

– *Twelve Types* (London: Arthur L. Humphreys, 1910).

Cowley, Malcolm (Ed.), *Writers at Work: The Paris Review Interviews* (New York: The Viking Press, 1964).

Dillard, Annie, *Pilgrim at Tinker Creek* (New York: Harper & Row Publishers, 1988).

Grozdanovitch, Denis, *La gloire des petites choses* (Paris: Bernard Grasset, 2022).

Guardini, Romano, *The Art of Praying: The Principles and Methods of Christian Prayer*, trans. by Prince Leopold of Loewenstein-Wertheim (Manchester, N.H., Sophia Institute Press, 1995).

– *Freedom, Grace and Destiny: An Interpretation of Existence,* trans. by John Murray, S.J. (Providence, R.I.: Cluny Media LLC, 2023).

– *Les signes sacrés*, trans. by Antoine Giraudet (Paris: Éditions Spes, 1938).

Haeffner, Gerd, "Gott der Gegenwart," in *Wege in die Freiheit. Philosophische Meditationen über das Menschsein* (Stuttgart: Kohlhammer, 2006), pp. 139-79.

– *In der Gegenwart leben. Auf der Spur eines Urphänomens* (Stuttgart: Kohlhammer, 1996).

– "Präsenz," in *Wege in die Freiheit. Philosophische Meditationen über das Menschsein* (Stuttgart: Kohlhammer, 2006), pp. 121-38.

– "Staunen," in *Wege in die Freiheit. Philosophische Meditationen über das Menschsein* (Stuttgart: Kohlhammer, 2006), pp. 11-25.

Haldas, Georges, "Les Sept Piliers de l'État de Poésie," in *Les Poètes malades de la Peste suivi de Les Sept Piliers de l'État de Poésie* (Lausanne: L'Age d'Homme, 2009).

Hartmann, Nicolai, *Ethics. Volume 2: Moral Values*, trans. by Stanton Coit (Atlantic Highlands, New Jersey: Humanities Press 1963).

Hegel, Georg Wilhelm Friedrich, *Aesthetics: Lectures on Fine Art*, trans. by T. M. Knox (Oxford: Oxford University Press, 1975).

Heyd, David, "Tact: Sense, Sensitivity, and Virtue," *Inquiry* 38 (1995), pp. 217-31.

Huxley, Aldous, "Knowledge and Understanding," in *Adonis and the Alphabet and Other Essays* (London: Chatto & Windus, 1956), pp. 39-72.

– "A Night at Pietramala," in *Along the Road: Notes and Essays of a Tourist* (London: Chatto & Windus, 1948), pp. 205-31.

Jankélévitch, Vladimir, *Liszt et la rapsodie. Essai sur la virtuosité* (Paris: Plon, 2019).

Kerényi, Karl, "Vom Wesen des Festes. Antike Religion und ethnologische Religionsforschung," in *Paideuma. Mitteilungen zur Kulturkunde*, Bd. 1, H. 2 (1938), pp. 59-74.

Kodály, Zoltán, *Mein Weg zur Musik. Fünf Gespräche mit Lutz Besch* (Zürich: Verlag der Arche, 1966).

Ladrière, Jean, "La ville, inducteur existentiel," in *Vie sociale et destinée* (Gembloux: Duculot, 1973), pp. 139–60.

Landsberg, Paul-Louis, *Problèmes du personnalisme* (Paris: Éditions du Seuil, 1952).

Lavelle, Louis, *The Dilemma of Narcissus*, trans. by W.T. Gairdner (London: George Allen & Unwin LTD, 1973).

– *La parole et l'écriture* (Paris: L'artisan du livre, 1947).

Leimeiszter, Barnabás, "Jelen lenni (To Be Present)," *Mandiner* 10, 3-11 (2023), p. 65.

Lewis, Clive Staples, *The Four Loves* (Glasgow: Collins, 1981).

– *Surprised by Joy* (San Francisco: HarperOne, 2017).

Lorenz, Konrad. *The Waning of Humaneness*, trans. by Robert Warren Kickert (Boston: Little, Brown and Company, 1987).

Marcel, Gabriel, "La musique et le règne de l'esprit," in *Présence de Gabriel Marcel*, Cahier 2-3, *L'esthétique musicale de Gabriel Marcel* (Paris: Aubier, 1980), pp. 49-59.

Matray, Bernard, *La présence et le respect. Éthique du soin et de l'accompagnement* (Paris: Desclée de Brouwer, 2004).

Mehl, Roger, "Structure philosophique de la Notion de Présence," *Revue d'histoire et de philosophie religieuse*, 38 (1958), pp. 171-6.

Merleau-Ponty, Maurice, *The Structure of Behavior*, trans. by Alden L. Fisher (Boston: Beacon Press, 1963).

Minkowski, Eugène, *Lived Time: Phenomenological and Psychopathological Studies*, trans. by Nancy Metzel (Evanston: Northwestern University Press, 1970).

– *Vers une cosmologie. Fragments philosophiques* (Paris: Éditions des compagnons d'humanité, 2022).

Morris, Paul Duncan, *Vladimir Nabokov: Poetry and the Lyric Voice* (Toronto: University of Toronto Press, 2011).

Morrow Lindbergh, Anne, *Gift from the Sea* (New York: Vintage Books, 1978).

Munch, Charles, *I am a Conductor*, trans. by Leonard Burkat (New York: Oxford University Press, 1955).

Nabokov, Vladimir, "The Creative Writer," in *Think, Write, Speak. Uncollected Essays, Reviews, Interviews and Letters to the Editor*, ed. by Brian Boyd and Anastasia Tolstoy (New York: Alfred A. Knopf, 2019), pp. 191-202.

– "Pushkin, or the True and Seemingly True (Essay 1937)," in *Think, Write, Speak. Uncollected Essays, Reviews, Interviews and Letters to the Editor*, ed. by Brian Boyd and Anastasia Tolstoy (New York: Alfred A. Knopf, 2019), pp. 118-32.

Percy, Walker, *The Moviegoer* (New York: Farrah, Straus and Giroux, 2019).

Picard, Max, *The World of Silence*, trans. by Stanley Godman (Washington, D.C., Gateway Edition, 1988).

– *Zerstörte und Unzerstörbare Welt* (Erlenbach-Zürich: Eugen Rentsch Verlag, 1951).

Pieper, Josef, "Music and Silence," in *Only the Lover Sings: Art and Contemplation*, trans. by Lothar Krauth (San Fransisco: Ignatius Press, 1990), pp. 53-6.

– "*Vita contemplativa* – The Contemplative Life," in *Only the Lover Sings: Art and Contemplation*, trans. by Lothar Krauth (San Fransisco: Ignatius Press, 1990), pp. 71-6.

Plessner, Helmuth, "Anthropologie der Sinne," in Hans-Georg Gadamer, Paul Vogler, ed., *Philosophische Anthropologie. Zweiter Teil* (Stuttgart: Georg Thieme Verlag, 1974), pp. 3-59.

Powys, John Cowper, *The Meaning of Culture* (London, Village Press. 1974).

Ricoeur, Paul, *La critique et la conviction. Entretien avec François Azouvi et Marc de Launay* (Paris: Fayard/Pluriel, 2010).

Rilke, Reiner Maria, *Notizen zur Melodie der Dinge* (Paris: Éditions Allia, 2018).

Romilly, Jacqueline de, *Ce que je crois* (Paris: Éditions de Fallois, 2012).

– "Diptyque," in *Honneur aux maîtres*, ed. by Marguerite Léna (Paris: Criterion, 1991), pp. 49-51.

Saint-Exupéry, Antoine de, *A Letter to a Hostage*, trans. by Jacqueline Gerst (London: Pushkin Press, 2008).

Sansot, Pierre, *Les gens de peu* (Paris: Presses universitaire de France, 1991), pp. 219-23.

Scheler, Max, "The Forms of Knowledge and Culture," in *Philosophical Perspectives*, trans. by Oscar A. Haac (Boston: Beacon Press, 1958), pp. 13–49.

Serres, Michel, *Corps* (Paris: Carnets Nord, 2017).

– *Le Tiers-Instruit* (Paris: Gallimard, 1992).

Simmel, Georg, "Bridge and Door," *Theory, Culture & Society* 11 (1994), pp. 5-10.

Storr, Anthony, *Music and the Mind* (New York: The Free Press, 1992).

Straus, Erwin Walter, *The Primary World of Senses: A Vindication of Sensory Experience*, trans. by Jacob Needleman (New York, Free Press of Glencoe, 1963).

Szerb, Antal, *The Third Tower: Journeys in Italy*, trans. by Len Rix (London, Pushkin Press, 2014).

Weischedel, Wilhelm, "Fragment über die Musik," in *Wirklichkeit und Wirklichkeiten. Aufsätze und Vorträge* (Berlin: De Gruyter, 1960), pp. 202-7.

Welte, Bernhard, "Meditation über Zeit," in *Zeit und Geheimnis. Philosophische Abhandlungen zur Sache Gottes in der Zeit der* Welt (Fribourg-en-Brisgau: Verlag Herder, 1975), pp. 15-27.

Wilder, Thornton, *Three Plays. Our Town, The Skin of Our Teeth, The Matchmaker* (New York: Harper & Brothers Publishers, 1958).

Wilfart, Serge, *Le chant de l'Être. Analyser, construire, harmoniser par la voix* (Paris: Albin Michel, 1994).

Williams John, *Butcher's Crossing* (New York: New York Review Books, 2007).

– *Stoner* (New York: New York Review Books, 2003).

Wittgenstein, Ludwig, *Philosophical Investigations*, ed. by Rush Rhees and G.E.M. Anscombe, trans. by G.E.M. Anscombe (Oxford: Blackwell, 2001).

INDEX

MIMESIS GROUP
www.mimesis-group.com

MIMESIS INTERNATIONAL
www.mimesisinternational.com
info@mimesisinternational.com

MIMESIS EDIZIONI
www.mimesisedizioni.it
mimesis@mimesisedizioni.it

ÉDITIONS MIMÉSIS
www.editionsmimesis.fr
info@editionsmimesis.fr

MIMESIS COMMUNICATION
www.mim-c.net

MIMESIS EU
www.mim-eu.com

www.ingramcontent.com/pod-product-compliance
Lightning Source LLC
LaVergne TN
LVHW092136071126
851857LV00038BB/504